Applications to Marine Disaster Prevention

AF213186

Naomi Kato
Editor

Applications to Marine Disaster Prevention

Spilled Oil and Gas Tracking Buoy System

Springer

Editor
Naomi Kato
Graduate School of Engineering
Osaka University
Suita, Japan

ISBN 978-4-431-56747-9 ISBN 978-4-431-55991-7 (eBook)
DOI 10.1007/978-4-431-55991-7

© Springer Japan 2017
Softcover reprint of the hardcover 1st edition 2016
This work is subject to copyright. All rights are reserved by the Publisher, whether the whole or part of the material is concerned, specifically the rights of translation, reprinting, reuse of illustrations, recitation, broadcasting, reproduction on microfilms or in any other physical way, and transmission or information storage and retrieval, electronic adaptation, computer software, or by similar or dissimilar methodology now known or hereafter developed.
The use of general descriptive names, registered names, trademarks, service marks, etc. in this publication does not imply, even in the absence of a specific statement, that such names are exempt from the relevant protective laws and regulations and therefore free for general use.
The publisher, the authors and the editors are safe to assume that the advice and information in this book are believed to be true and accurate at the date of publication. Neither the publisher nor the authors or the editors give a warranty, express or implied, with respect to the material contained herein or for any errors or omissions that may have been made.

Printed on acid-free paper

This Springer imprint is published by Springer Nature
The registered company is Springer Japan KK

Preface

This book deals with the results of the research project funded by a Grant-in-Aid for Scientific Research (S) of the Japan Society for the Promotion of Science (No. 23226017) from FY 2011 to FY 2015 with the title "A new spilled oil and gas tracking autonomous buoy system and application to marine disaster prevention system". The book was written because we thought that a book is the best way to describe the details of our views, methods, and results as a whole project. The research project was initiated with the *Deepwater Horizon* oil spill accident off Louisiana USA in the Gulf of Mexico in 2010, considered to be the largest in the history of the petroleum industry, following our past research on spilled oil tracking autonomous buoy systems to autonomously track spilled oil drifting on the sea surface and obtain useful data in real time.

This project was carried out by four groups:

Group 1 dealt with the development and experiments of an underwater autonomous robot equipped with sensors for measuring oceanographic data and dissolved gas and oil,
Group 2 dealt with the development and experiments of an autonomous surface vehicle equipped with sails, the size of which was adjustable, and a detecting sensor of oil slicks on the sea surface,
Group 3 dealt with simulation of the behavior of underwater oil and gas after a blowout from the seabed in deep water,
Group 4 dealt with simulation of drifting spilled oil with a data assimilation scheme.

The leader of Group 1 was Mr. Mahdi Choyekh from Tunisia. He started his study in Japan on April 1, 2011, as a research student at Osaka University and is now a doctoral student in our laboratory after finishing our master's course; he will finish the doctoral program in September 2016. He devoted himself extensively to the development and operation of the autonomous underwater robot *SOTAB-I* mainly from the aspect of its total software. Mr. Yasuaki Yamaguchi and Mr. Ryan Putra Dewantara, the authors of Chap. 3 and students in the master's degree course of our department, made great contributions to the research of Group 1.

I would like to express my sincere gratitude to Prof. Hajime Chiba of the National Institute of Technology, Toyama College, as one of our cooperating researchers for contributing great efforts to carry out many field experiments of *SOTAB-I* in Toyama Bay and off Joetsu. I would like to thank Dr. Toshinari Tanaka and Dr. Muneo Yoshie of the Port and Airport Research Institute (PARI) as our cooperating researchers who gave us the opportunities to carry out water tank tests of *SOTAB-I* in PARI. I am indebted to Prof. Eiji Kobayashi of Kobe University as one of our cooperating researchers for giving us the opportunity to use the training ship *Fukaemaru* to carry out a field experiment of *SOTAB-I*. I am also indebted to Dr. Timothy Short of SRI International, who dealt with underwater mass spectrometry in installing it inside *SOTAB-II* and in acquiring and analyzing the data. I am thankful to Prof. Masakazu Arima and Prof. Norimitu Sakagami for their contributions to the field experiments of *SOTAB-I*. The research of Group 1 is greatly indebted for the contributions by Mr. Ryota Kimura, Mr. Masahiro Ukita, Mr. Tsukasa Kita, Ms. Mizuho Fujihara, and Ms. Keiko Kamezuka, former students in the Kato laboratory.

The leader of Group 2 was Mr. Swarn Singh Rathour from India. He started his study in Japan on April 1, 2013, as a research student in Osaka University, is now a doctoral student in our laboratory, and will finish his doctorate in September 2016. He contributed much to the operation of the autonomous underwater robot *SOTAB-II* mainly from the aspect of its control system. Mr. Naoto Tanabe, a master's degree student in our department, made a contribution to the research of Group 2. I would like to thank Dr. Toshinari Tanaka and Dr. Muneo Yoshie of PARI for supporting the field test of the older version of *SOTAB-II*. I am indebted to Prof. Emeritus Yutaka Masuyama of the Kanazawa Institute of Technology for giving us the offset data of the yacht *KIT34*, which he designed, for the design of the new *SOTAB-II*. The research of Group 1 owes much to the contributions by Mr. Lubin Yu, Mr. Tatsuya Akamatsu, and Ms. Yukino Hirai, former students in the Kato laboratory.

The leader of Group 3 was Assistant Prof. Yohei Takagi under the guidance of Prof. Yasunori Okano. They are members of the Department of Chemical Engineering, Graduate School of Engineering Science, Osaka University. I am thankful to Assistant Prof. Yohei Takagi for his great efforts to enthusiastically guide the students in this group. The research of Group 3 owes much to the contributions by Mr. Satoru Kunikane and Mr. Sho Kawahara, former students in the Kato laboratory, and to Mr. Tatsumi Iwao, a former student in the Okano laboratory.

The leader of Group 4 was Associate Professor Hiroyoshi Suzuki of the Department of Naval Architecture and Ocean Engineering, Graduate School of Engineering, Osaka University. The research of Group 4 owes much to the contributions by Mr. Kosuke Iwai and Mr. Takeaki Tsutsukawa, former students in the Kato laboratory.

I would like to thank Assistant Professor Hidetaka Senga for supporting our research, especially for Group 1 and Group 2.

I would also like to thank all those who have helped in many ways in the progress of our research, especially Mr. Yuzuru Ito of Ocean Engineering Research, Inc., Mr. Toyohiko Sakai of SKE, Inc., Mr. Hiromichi Ito of Technical Service, Inc., staff members of *R/V Pelican* of LUMCON, Mr. Tadashi Morita of W2 Inc., the staff of

Fukaemaru of Kobe University, and the staffs of *Wakashiomaru* and *Sazanami* of the National Institute of Technology, Toyama College.

I am appreciative of our secretary, Ms. Mamiko Eguchi, for supporting our research from the aspect of financial management and for editing the manuscripts of this book.

Finally, I would like to thank the Japan Society for the Promotion of Science for financially supporting our research for 5 years.

Suita, Japan Naomi Kato

Contents

Chapter 1
Introduction

Naomi Kato

Abstract The first part of this chapter describes the background, objectives, and outline of the spilled oil tracking autonomous buoy (SOTAB) project funded by the Japan Society for the Promotion of Science for 5 years since fiscal year 2011. The latter part explains the outlines of the subsequent seven chapters.

Keywords SOTAB project • Background • Objectives • Outline • Chapters

There have been many major marine oil spills in recent years. These spills damage not only the ocean environment but also regional economies. Once spilled oil washes ashore, it is difficult to recover it effectively. This results in a high residual amount of spilled oil and long-term damage to the environment as well as to marine and human life. In January 1997 in the Sea of Japan, Russian tanker Nakhodka broke into two sections over 100 km from the coast, which resulted in the heavy oil spill of 6240 kl and polluted a long stretch of the Sea of Japan coast (Moller 1997). The explosion of the offshore oil rig Deepwater Horizon off Louisiana in the Gulf of Mexico in 2010 and the oil spill from the seabed at the depth of 1552 m, amounting to about 0.78 million kl, drew our attention to the danger of a large oil spill from subsea oil production systems (Elfring and Waddell 2014). On the other hand, once gas leaks from the seabed by accident from a subsea oil production system or due to seismic activity and subsea landslide in the area of ample reserves of methane hydrate in the sea, it seriously damages not only offshore platforms but also the natural environment. In March 2012, a gas leak occurred at the Elgin offshore gas production platform in the North Sea from a point some 4000 m below the seabed into a well bore and continued over 7 weeks. As methane gas was released, ships and airplanes were barred from a wide area (Wilson et al. 2013).

To prevent such oil spills and gas leaks from spreading and causing further damage over time and space, early detection and monitoring systems should be deployed around the offshore oil and gas production system. If an accidental oil

---

N. Kato (✉)
Department of Naval Architecture and Ocean Engineering, Osaka University, 2-1, Yamada-oka, Suita 565-0871, Osaka, Japan
e-mail: kato@naoe.eng.osaka-u.ac.jp

© Springer Japan 2017
N. Kato (ed.), *Applications to Marine Disaster Prevention*,
DOI 10.1007/978-4-431-55991-7_1

spill occurs, the exact location of the drifting oil and the meteorological and oceanographic data should be collected in real time so that oil recovery operations can be smoothly coordinated and adequate measures such as deployment of oil recovery machines can be taken at coastal areas using information obtained from the oil drifting simulation enhanced by data assimilation to precisely predict where the spilled oil will wash ashore.

From a scientific viewpoint, apart from the abovementioned social background, some challenging unsolved problems remain in this field, such as multi-scale phenomena including the composition and decomposition of deep-sea methane hydrates and the advection and diffusion of oil and gas, multi-physical phenomena including the thermal dynamics of gas composition and decomposition and the hydrodynamics of the behavior of oil and gas, and environmental forecasting in a precise manner using assimilation based on real-time monitored data at sea.

There are two types of underwater robots autonomously monitoring the marine environments in 3-D space from sea surface to seabed over the long term. One is the Argo float (Roemmich et al. 2009), and the other is the underwater glider (Eriksen et al. 2001). Argo float floats vertically and repeats descending and ascending in the vertical direction. However, it does not have the function of active movement in the horizontal direction. Underwater glider has a streamlined body with fixed wings. It can descend and ascend using a buoyancy control device, while moving in the horizontal plane like a glider for long distances. However, the ratio of vertical movement distance to horizontal movement distance is small.

There are some methods to deal with the drifting spilled oil on the sea surface. Drifting buoys are used to track the oil spill (Goodman et al. 1995). The disadvantage of these drifting buoys is that once they are apart from the oil slick, they are unable to track it. If the sea condition is fine and safe enough, the vessels can track an oil slick using the X-band radar detection method (Miros AS). A plane can fly over the sea and detect the oil slick. However, it is difficult to detect the spilled oil from the air at night. Even though aircraft can apply a fluorescence lidar system (Yamagishi et al. 2000) to detect oil slicks at night, the limitation of its endurance can render it impossible to track the oil slick continuously. Satellite remote sensing is available to detect the location of an oil spill (Jensen et al. 2008). However, this method is not used more frequently than the aircraft method.

To overcome the weak points of existing oil spill monitoring methods and, moreover, to deal with gas and oil blowout accidents, a project funded by the Japan Society for the Promotion of Science has being carried out for 5 years since FY2011. This system consists of the spilled oil tracking autonomous buoy (SOTAB) and the land base. Figure 1.1 depicts the concept of this system. The underwater autonomous robot named SOTAB-I observes 3-D distributions of gas and oil blowing out from the seabed (Measure 1). On the sea surface, multiple autonomous buoys with controllable sails, termed SOTAB-II, observe the oil slick (Measure 2). One of the important roles of SOTAB-I and SOTAB-II is to send the useful real-time data around it to the land base or mother ship. Simulations for predicting the diffusion and drifting of spilled oil and gas are carried out at the land base by incorporating the real-time data from these robots (Measure 3). The

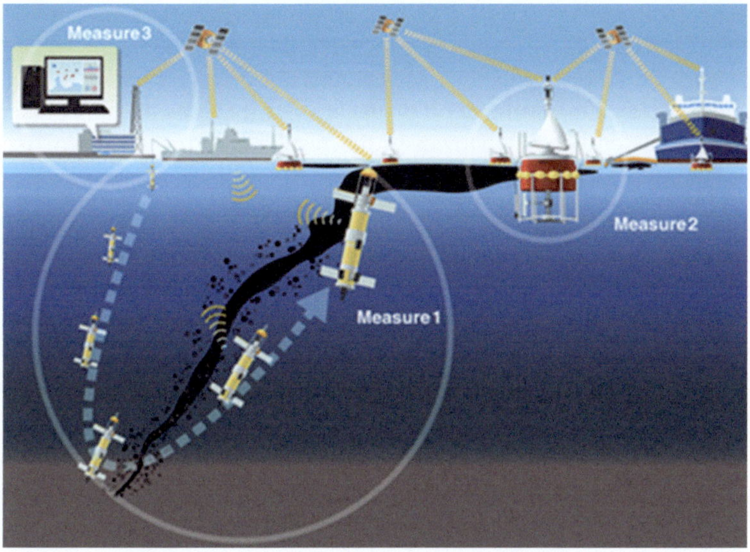

Fig. 1.1 Concept of spilled oil and gas tracking autonomous system

accuracy of this forecast is enhanced by data assimilation using the real-time data from the SOTABs. Then, adequate measures can be taken at coastal areas where the oil slick will drift ashore. Hereafter, this project is referred to as "SOTAB project."

Before this project, our team had been developing two types of SOTAB to track spilled oil autonomously and to obtain useful data in real time (Senga et al. 2009, 2011).

The first type of SOTAB, named SOTAB-I, could descend and ascend underwater to guide itself to spilled oil using two oil-detecting sensors consisting of a contact sensor focused on the viscosity of the spilled oil and a noncontact sensor using four charge-coupled device (CCD) cameras and global positioning system (GPS) and an anemometer. For the autonomous tracking mechanism, the buoy is equipped with a buoyancy control device (BCD) and two pairs of rotational fins. It detects and tracks the spilled oil autonomously using the above two sensors, the BCD, and the rotational fins as follows. When the SOTAB-I drifts on the sea surface, the contact sensor installed on top senses the fluctuation of the viscosity of the surrounding fluid. If the contact sensor detects no spilled oil, the buoy starts to descend by reducing its buoyancy. After it reaches the target depth, the noncontact sensor captures images of the sea surface and detects the direction and horizontal distance to the spilled oil. The SOTAB-I then starts to ascend toward the detected spilled oil by increasing its buoyancy and adjusting the angles of its fins. At the sea surface, the contact sensor again senses the viscosity of the surroundings. If the SOTAB-I has not successfully entered the area where oil has spilled, these descending and ascending procedures are repeated. This tracking procedure was adopted to conserve the electric power of the SOTAB for long-term observation from the beginning of an oil spill. Field experiments such as data measurements and autonomous tracking experiments were carried out. Here, neoprene rubber sheets were used as the target on the sea surface instead of actual heavy oil. SOTAB-I successfully detected the artificial oil on the sea surface using the noncontact sensor and tracked it autonomously. However, the results also revealed two issues that needed to be addressed:

(a) During the tracking procedure, the drift of the SOTAB-I was strongly affected by the water current, while that of the rubber sheet floating on the sea surface was strongly affected by the wind. This same wind effect occurs with oil spills. As a result, the SOTAB-I occasionally floated away from the rubber sheet, especially when the wind and water current directions differed.
(b) Occasionally, the noncontact sensor could not precisely detect the rubber sheet because of the low transparency of seawater or because of obstacles.

An oil spill's direction of drift is greatly affected by the wind. In order to effectively gain thrust force from the wind and drift along with the spilled oil, a new SOTAB, named SOTAB-II, was constructed with a sail. The direction of this sail and its size are both adjustable. In drifting mode, the SOTAB-II drifts along with the spilled oil by controlling the size and direction of this sail as circumstances change. A cylindrical shape was adopted for the hull for the SOTAB-II to rapidly respond to changes in the drifting direction of the oil slick. The oil-detecting sensor was attached to the top of the SOTAB-II so it could detect spilled oil above the sea surface, regardless of whether or not the seawater is transparent. Meanwhile, an auxiliary propulsion mechanism was installed on the SOTAB-II. A normal underwater thruster would require two motors: one for the thruster and the other for azimuth control. Therefore, we planned to use an oscillating fin as a propulsor, like the caudal fin of a fish. The oscillating fin makes it possible to control the advancing speed and direction by one motor. This propulsion mechanism is mainly used to control the drift of the SOTAB-II with orders from the land base. The main autonomous tracking mechanism of the SOTAB-II is accomplished by controlling the direction and size of the sail. When the SOTAB-II is separated from the spilled oil by a gust of wind, it can detect and track the spilled oil autonomously by using the oil-detecting sensor and the propulsion mechanism. However, to conserve electric power, this actively autonomous tracking procedure using the auxiliary propulsion mechanism must stay off. The drifting performance of the SOTAB-II was verified with several sea trials. The results revealed that the SOTAB-II can drift not only along water currents but also using the wind effect. As only the sail size was changed and tested in the sea trials, the drifting direction of SOTAB-II did not become closer to that of the spilled oil.

Simulation studies on drifting spilled oil have been performed by many researchers, including using data assimilation from satellites and drifting buoys to increase the precision of predictions of the simulation. However, studies on the utilization of real-time data on the location of drifting spilled oil and surrounding meteorological and oceanographic data for the simulation have not been performed yet. Studies on the rising and drifting behavior of oil and gas seepage from deepwater seabed, including hydrothermal chemical reactions, have been performed extensively. However, studies on the assimilation of data on the diffusion of plumes of oil and gas and surrounding oceanographic data to increase the precision of simulations of the rising and drifting behavior of oil and gas seepage from deepwater seabed have not been performed yet.

From the abovementioned experiences, SOTAB project was set up to develop a new type of autonomous underwater vehicle, SOTAB-I, for the autonomous tracking and monitoring of spilled plumes of oil and gas from subsea production facilities as the first case of autonomous underwater vehicle. SOTAB project was also established to develop a new type of autonomous surface vehicle (ASV), SOTAB-II, for real-time tracking and monitoring of oil spill spread and drift using an oil sensor. Furthermore, to improve the precision of predictions of oil and gas behavior using simulation models by incorporating real-time measured data from the buoy robots SOTAB-I and SOTAB-II, a data fusion method incorporating real-time measured data from buoy robots in the simulation models for not only gas and oil blowouts but also spilled oil drifting on the sea surface will be developed.

This book begins with Chap. 2 dealing with lessons from marine-based oil spill and gas leak accidents, including major marine-based oil spill and gas leak accidents, the subsea environmental effects by the Deepwater Horizon oil spill accident, surface oil slick behavior after the Deepwater Horizon oil spill accident, and the impact of the Deepwater Horizon oil spill accident on contingency plans, preparedness, and regulations. Chapter 3 deals with the development and operation of the underwater robot SOTAB-I in SOTAB project for the autonomous tracking and monitoring of subsea plumes after oil spill and gas leak from the seabed and analyses of the measured data. The experimental results in the shallow water of the Gulf of Mexico in the USA demonstrated the ability of SOTAB-I to collect the dissolutions of substances such as hydrocarbons in seawater. Deepwater experiments were also conducted in Toyama Bay, Japan, and the enabled demonstration of the ability of SOTAB-I to establish the vertical water column distribution of oceanographic data, such as temperature, salinity, and a high-resolution profile of water currents. Chapter 4 describes an autonomous robotic platform propelled by wind and water currents for long-term monitoring of spilled oil on the ocean surface. This chapter delineates the design of the oil spill-tracking ASV. This chapter also describes a sensor-based guidance, navigation, and control system for oil spill tracking by an ASV in unsteady and uncertain environments. The primary contribution of this chapter is to propose a clustering-based decision-making algorithm for sailing the ASV based on a complete time history of the scanned area around the vehicle by the oil-detecting sensor. Field test experiments are used to validate the operation of the ASV with respect to oil spill tracking using the guidance, navigation, and control systems based on onboard sensor data for tracking artificial oil targets.

This book moves on to discuss simulations of the diffusion and drifting of spilled oil and gas. Chapter 5 describes the development of two numerical models for tracking oil and gas blowout from the seabed in deep water based on conventional plume or Lagrangian approaches. The simulations are carried out for the Deep Spill experiment in the Norwegian Sea (Johansen et al. 2001), and the two test cases of oil and gas discharges are calculated. The two numerical models for tracking oil and gas blowout in deep water are developed based on the conventional plume or Lagrangian approaches. One is a simple oil tracking model that can be used if droplet breakup and coalescence are not necessary. The other involves the tracking of methane that requires various physical models and the kinetics of methane hydrate. Chapter 6 deals with multiphase flow simulation for oil and gas blowout in water to investigate the effect of liquid–gas interaction on plume structure. The mass and momentum balance equations are solved for each phase, and the interactions are considered with the Euler–Euler approach. To consider the interphase interaction between two dispersed phases in the oil–gas–water three-phase system, the drag force between oil and gas in the momentum equation is applied. Chapter 7 describes spilled oil drifting simulation using the 1997 Nakhodka oil spill in the Sea of Japan as a test case for the simulation. To improve the accuracy of the simulation, a data assimilation (DA) scheme is applied for the atmospheric simulation using observations. Before applying the DA scheme to the oil drifting simulation, numerical experiments are carried out to select the DA scheme. Subsequently, the efficacy of the DA when data from the SOTAB-II are supposed to be applied is investigated to forecast the phenomena of oil drifting on the sea surface.

The last chapter summarizes the contents from Chaps. 2, 3, 4, 5, 6, and 7.

References

Elfring C, Waddell K (2014) Research to advance health, environment, and oil system safety in the gulf of Mexico and other coastal regions: an update from the National Academy of Sciences. In: Proceedings of international oil spill conference 2014, vol 2014 issue 1. Savannah, 5–8 May 2014, pp 340–347

Eriksen CC et al (2001) Seaglider: a long-range autonomous underwater vehicle for oceanographic research. IEEE J Ocean Eng 26(4):424–436

Goodman RH, Simecek-Beatty D et al (1995) Tracking buoys for oil spills. In: Proceedings of international oil spill conference 1995, vol 1995 no. 1. Savannah, 27 Feb–2 Mar 1995, pp 3–8

Jensen HV et al (2008) Recent experience from multiple remote sensing and monitoring to improvement oil spill response operations. International oil spill conference 2008, vol. 2008 issue. 1. Savannah, 4–8 May 2008, pp 407–412

Johansen Ø et al (2001) DeepSpill JIP – experimental discharge of gas and oil at Helland Hansen, Jun 2000, Technical report. SINTEF Report STF66 F01082, SINTEF applied chemistry, Trondheim

Miros AS. http://www.miros.no/download/osd/DB-26_01_DataSheet_MirosOSD.pdf

Moller TH (1997) The NAKHODKA oil spill response-the technical adviser's perspective. PAJ oil spill symposium 1997 – lessons learnt from the Nakhodka incident. In: Proceedings of PAJ oil spill symposium'97, Tokyo, 10–11 July 1997

Roemmich D et al (2009) The Argo program observing the global ocean with pro-filing floats. Oceanography 22(2):34–43. doi:10.5670/oceanog.2009.36

Senga H et al (2009) Spilled oil tracking autonomous buoy system. Adv Robot 23(9):1103–1129. doi:10.1163/156855309X452476

Senga H et al (2011) Development of a new spilled oil tracking autonomous buoy. MTS J 45(2):43–51, http://dx.doi.org/10.4031/MTSJ.45.2.8

Wilson G et al (2013) Following the safe evacuation of all offshore personnel . . . the environmental response takes primacy: learning from the Elgin G4 well major incident. In: Proceedings of SPE offshore Europe oil and gas conference and exhibition 2013, Aberdeen, 3–6 Sept 2013
Yamagishi S et al (2000) Development and test of a compact lidar for detection of oil spills in water. In: Proceedings of SPIE 4154 hyperspectral remote sensing of the ocean, Sendai, 9 Oct 2000, pp 136–144

Chapter 2
Lessons from Marine-Based Oil Spill and Gas Leak Accidents

Naomi Kato

Abstract This chapter first describes past major marine-based oil spill and gas leak accidents around the world. The Deepwater Horizon (DWH) oil spill that occurred on April 20, 2010, in the Gulf of Mexico (GOM) is considered the largest in the history of the petroleum industry. Second, the investigations into the subsea environmental effects by the DWH oil spill accident are described from the viewpoints of tools, structure of subsurface plumes, and the effect of ocean circulation in the GOM on subsea hydrocarbon compound drifting. Third, surface oil slick behavior after the DWH oil spill accident is discussed from the viewpoints of satellite remote sensing and the effect of the GOM loop current. Finally, the impact of the DWH oil spill accident on contingency plan, preparedness, and regulations is explained.

Keywords Oil spill • Gas leak • Subsurface plume • Ocean circulation • Surface oil slicks • Impacts • Deepwater Horizon

2.1 Major Marine-Based Oil Spill and Gas Leak Accidents

The DWH oil spill occurred on April 20, 2010, in the GOM, about 66 km off the southeast coast of Louisiana at a water depth of about 1522 m, on the BP-owned Transocean-operated Macondo Prospect. After the DWH oil rig exploded and sank, oil spill from the seabed continued for 87 days until it was capped on July 15, 2010. The US Government estimated the total discharge at 0.57 million kl (National Commission on the BP Deepwater Horizon Oil Spill and Offshore Drilling 2011). This oil spill accident is considered the largest in the history of the petroleum industry; the second largest was Ixtoc I (Atwood 1980) in the southwestern GOM on June 3, 1979, which was operated by Mexico's government-owned oil company Pemex and released an estimated 3.3 million m^3 of oil for nearly 10 months.

N. Kato (✉)
Department of Naval Architecture and Ocean Engineering, Osaka University, 2-1, Yamada-oka, Suita 565-0871, Osaka, Japan
e-mail: kato@naoe.eng.osaka-u.ac.jp

© Springer Japan 2017

N. Kato (ed.), *Applications to Marine Disaster Prevention*,
DOI 10.1007/978-4-431-55991-7_2

The Spanish oil tanker Castillo de Bellver, carrying 250,000 t of light crude oil from the Persian Gulf to Spain, caught fire off Saldanha Bay, approximately 70 miles northwest of Cape Town, South Africa, on August 6, 1983. A total of 0.297 million m^3 oil was spilled on the sea surface (Read 2011). This oil spill accident is the largest among oil tanker oil spill accidents in the world. Japan experienced a large-scale oil spill accident on January 2, 1997; the Russian tanker Nakhodka, which was 26 years old, had broken in half and sunk in the Sea of Japan on the way from Shanghai to Petropavlovsk; it had carried 19,000 kl C crude oil. The main part of the tanker sank to the seafloor 2500 m below the surface and its bow drifted ashore. About 6240 kl oil was spilled on the sea surface and 12,500 kl sank to the seafloor (Moller 1997).

The recent gas leak accident occurred on March 25, 2012, at the wellhead platform on the Elgin gas field, which is in the UK North Sea approximately 240 km east of Aberdeen and is operated by the French energy group Total. The gas leak resulted in the shutdown of production and evacuation of personnel. The leak continued for over 7 weeks, and it was stopped after well intervention work on May 16, 2012 (Wilson et al. 2013). On February 18, 2015, a condensate leak occurred on the Statoil-operated Gudrun platform in the Norwegian part of the North Sea. Condensate leaked from the pipeline and resulted in gas distribution. The volume of the leak from the condensate was estimated at 4 kl, representing a major incident potential if ignited. The leak was recorded by the gas detectors, and ignition sources were disconnected and the deluge system was operated automatically (OGJ editors 2015). Piper Alpha was a North Sea oil production platform operated by Occidental Petroleum (Caledonia) Ltd. The platform began production in 1976 first as an oil platform, and then it was later converted to gas production. An explosion and the resulting oil and gas fires destroyed it on July 6, 1988, killing 167 men, with only 61 survivors (Mcginty 2008). At the time of the disaster, the platform was responsible for approximately 10 % of North Sea oil and gas production and was the worst offshore oil disaster in terms of lives lost and industry impact.

2.2 Subsea Environmental Effects by the DWH Oil Spill Accident

After the DWH oil spill accident, many experimental surveys of subsea environments were carried out using research vessels such as remotely operated vehicles (ROVs) and autonomous underwater vehicles (AUVs). Camilli et al. (2010) used an AUV and a cable-lowered sample collection rosette to record mass spectrometric and fluorescence data during vertical profiling using the ship's sampling rosette approximately 4 km from the leak source and confirmed the presence of a large aromatic hydrocarbon plume at 1000- to 1200-m depth, as well as a more diffuse plume between 50- and 500-m depth. Ryan et al. (2011) coordinated ship and AUV operations to locate and study deep hydrocarbon. The ship hydrocast survey localized the maximum optical signals of a deep plume, centered at 1150-m depth, approximately 13 km southwest of the blowout. The AUV conducted a high-resolution survey of the plume structure at this position using a colored dissolved organic matter (CDOM) fluorometer, which indicated small-scale topographic influences on plume transport.

The DWH oil well blowout released large quantities of oil and gaseous hydrocarbons such as methane into the deep ocean. Ryerson et al. (2012) measured initial hydrocarbon compositions along different transport routes to the deep subsurface plumes, to the surface slick, and to the atmosphere during the DWH oil spill. They reported that the primarily soluble mixture detected in subsurface plumes made up about 35 % of the leaking mass and that the insoluble, nonvolatile mixture detected in the surface slick made up about 10 % of the leaking mass. Reddy et al. (2012) collected and analyzed the end-member sample of gas and oil that flowed from DWH. They found that the material had a gas-to-oil ratio (GOR, defined as the standard cubic feet per petroleum barrel at 15.6 °C and 1 bar) of 1600 standard cubic feet per petroleum barrel. They estimated the total C_1–C_5 hydrocarbons (methane, ethane, propane, butane, pentane) released to the water column as 1.7×10^{11} g using the federally estimated net liquid oil release of 0.48 million kl. Their results showed that most of the C_1–C_3 hydrocarbons and a significant fraction of water-soluble aromatic compounds were retained in the deepwater column, whereas the relatively insoluble petroleum components were predominantly transported to the sea surface or deposited on the seafloor. Joye et al. (2011) revealed layers of dissolved hydrocarbon gases at between 1000- and 1300-m depth, where concentrations exceeded background levels by up to 75,000 times. They suggested that microbial consumption of these gases could lead to large-scale and long-term exhaustion of oxygen in the hydrocarbon-enriched waters. Kessler et al. (2011) discussed the effect of the spilled methane from the DWH wellhead on the oxygen distribution in deep water. Based on methane and oxygen distributions measured at 207 stations throughout the affected region, they found that a large quantity of oxygen was respired within about 120 days from the onset of release, primarily by methanotrophs, and a residual microbial community containing methanotrophic bacteria was formed. White et al. (2012) found that coral colonies at one site 11 km southwest of the Macondo well showed widespread signs of stress and strong evidence that flocculent material associated with corals contained oil from the Macondo well.

There was a possibility that some portion of the hydrocarbons issuing from the DWH wellhead would remain in subsurface water to be carried by the prevailing ocean circulation. Weiberg et al. (2011) applied existing nowcast/forecast models for the ocean circulation of the eastern GOM to respond to a tracking problem not only for the oil that reached the surface but also for the hydrocarbon compounds that remained at depth. By assuming that some compounds would reach certain levels and they would be carried three dimensionally by the currents there, they deployed virtual drifters at several depths and the surface, and to mimic the continual flow of oil, they added new virtual particles. Their results were in qualitative agreement with the limited observations and inferences of subsurface hydrocarbon locations obtained by ship surveys. Noth et al. (2011) used an analytical multiphase plume model combined with time-varying flow and hydrographic fields generated by a 3-D hydrodynamic model as input to a Lagrangian transport model to simulate the transport of oil droplets dispersed at depth from the DWH oil spill. They reported that the plume model predicted a stratification-dominated near field, where small oil droplets were trapped by density stratification. Their simulated droplet trajectories showed that droplets with diameters between 10 and 50 μm formed a distinct subsurface plume, which was transported horizontally and remained on the subsurface for >1 month. In contrast, droplets with diameter $\geq$90 μm rose rapidly to the surface.

A variety of techniques was used for oil spill response after the DWH oil spill accident, including containment of oil on the surface, dispersal, and removal. The critical difference in the case of the DWH oil spill accident from that of oil spill accidents of oil tankers was that oil spill and gas leak occurred and continued from the deep seabed. For the dispersal of the spilled oil, 7000 kl dispersants were used; of this 2920 kl were used subsea at the wellhead. The first subsea injection of dispersant directly into the oil at the source was carried out. Mixing the dispersants with the oil at the wellhead is said to keep some oil below the surface and allow microbes to digest the oil before it reaches the surface. However, Kleindienst et al. (2015) showed the evidence that chemical dispersants applied to either deep water or surface water from the Gulf of Mexico did not stimulate oil biodegradation. Direct measurement of alkane and aromatic hydrocarbon oxidation rates revealed either suppression or no stimulation of oil biodegradation in the presence of dispersants.

2.3 Surface Oil Slick Behavior After the DWH Oil Spill Accident

To detect and map the surface oil slicks after the DWH oil spill accident, airborne and spaceborne passive and active remote sensing has been extensively applied. Leifer et al. (2012) explained that state-of-the-art satellite and airborne marine oil spill remote sensing applied to the DWH oil spill. Oil slick thickness and oil-to-water emulsion ratios were derived quantitatively for thick (>0.1 mm) slicks from AVIRIS (airborne visible/infrared imaging spectrometer) data using a spectral library approach based on the shape and depth of near-infrared spectral absorption features (Clark et al. 2010). MODIS (moderate resolution imaging spectroradiometer) satellite, visible-spectrum broadband data of surface slick modulation of sunglint reflection, was used for extrapolation to the total slick. A multispectral (visible and thermal) expert system was developed using a neural network approach to provide rapid image analysis for oil spill thickness estimation. Synthetic aperture radar (SAR) data from airborne and satellite (SSAR) platforms were used widely for surface wave detection. However, SAR generally cannot discriminate thick (>100 μm) oil slicks from thin sheens (up to 0.1 μm). On the other hand, UAVSAR (unmanned aerial vehicle SAR) has significantly greater signal-to-noise ratio and finer spatial resolution. These characteristics led to successful pattern discrimination related to the combination of oil slick thickness, fractional surface coverage, and emulsification at the DWH oil spill accident (Jones et al. 2011).

The ocean circulation patterns of the GOM loop current (LC) system and their effects on the advection of the oil discharged during the DWH accident were studied using in situ surface drifter trajectories and satellite observations (Liu et al. 2011). These observations included altimetry-derived surface geostrophic velocities, sea surface temperature, ocean color, and surface oil locations. They found that whereas the LC usually may extend as far north as the northern GOM shelf slope, the LC remained south of where the main body of surface oil was located throughout the DWH event, and that only a small portion of the surface oil was entrained into the LC system. Walker et al. (2011) employed satellite data in tandem with in situ current and wind measurements to track surface oil and to understand the reasons for the observed large-scale motions during the 84 days. They revealed the merging of three cyclonic eddies along the LC's northern margin, finally forming a larger and stronger cyclonic eddy. They concluded that this larger cyclonic eddy controlled the motion of the oil/dispersant mixture into deep water, tripling the area of surface oiling.

2.4 Impact of the DWH Oil Spill Accident on Contingency Plan, Preparedness, and Regulations

After the DWH oil spill accident, the National Commission on the Deepwater Horizon Oil Spill and Offshore Drilling (2011) was created as an independent and nonpartisan entity by the US President to determine the causes of the disaster and to improve the country's ability to respond to spills and to recommend reforms to make offshore energy production safer. The commission presented the following seven recommendations for addressing the causes and consequences of the spill with a focus on the government's role:

(A) Improving the safety of offshore operations
(B) Safeguarding the environment
(C) Strengthening oil spill response, planning, and capacity
(D) Advancing well-containment capabilities
(E) Overcoming the impacts of the DWH spill and restoring the GOM
(F) Ensuring financial responsibility
(G) Promoting congressional engagement to ensure responsible offshore drilling

The first recommendation includes (1) the need for a new approach to risk assessment and management and (2) the need for a new, independent agency with the primary responsibility of regulating the offshore oil and gas industry. The third recommendation includes (1) the need for improved oil spill response planning, (2) the need for a new approach to handling spills of national significance, (3) the need to strengthen state and local involvement, (4) the need for increased research and development to improve oil spill response, and (5) the need for new regulations to govern the use of dispersants.

Major offshore oil producers, the USA, the UK, and Norway, reviewed and evaluated their regulatory approaches after the DWH oil spill accident (Bennear 2015). In the US approach, regulation has been set up through command-and-control technology standards governing specific safety systems. On the other hand, the European approach, led by the UK and Norway, has adopted a more risk-based regime where the responsibility of safety is shared among the industry, workers, and government. The reviews and evaluations of the DWH oil spill accident have resulted in a convergence of approaches in all three countries to a system that combines a strict liability regime for operators with risk-based safety and environmental management systems and command-and-control regulations.

References

Atwood DK (1980) Proceedings of a symposium on preliminary results from the September 1979 researcher/pierce Ixtoc-I cruise. Key Biscayne. 9–10 June 1980

Bennear LS (2015) Offshore oil and gas drilling: a review of regulatory regimes in the United States, United Kingdom, and Norway. Rev Environ Econ Policy 9(1):2–22. doi:10.1093/reep/reu013

Camilli R et al (2010) Tracking hydrocarbon plume transport and biodegradation at Deepwater Horizon. Science 330(6001):201–204. doi:10.1126/science.1195223

Clark RN et al (2010) A method for quantitative mapping of thick oil spills using imaging spectroscopy. U.S. Geological Survey open-file report number 2010–1167 (p 51)

Jones CE et al (2011) Studies of the Deepwater Horizon oil spill with the UAVSAR radar. In: Liu Y et al (eds) Monitoring and modeling the Deepwater Horizon oil spill: a record-breaking enterprise. AGU, Washington, pp 33–50

Joye SB et al (2011) Magnitude and oxidation potential of hydrocarbon gases released from the BP oil well blowout. Nat Geosci 4:160–164. doi:10.1038/ngeo1067

Kessler JD et al (2011) A Persistent oxygen anomaly reveals the fate of spilled methane in the Deep Gulf of Mexico. Science 331(6015):312–315. doi:10.1126/science.1199697

Kleindienst S et al (2015) Chemical dispersants can suppress the activity of natural oil-degrading microorganisms. PNAS 112(48):14900–14905, www.pnas.org/cgi/doi/10.1073/pnas.1507380112

Leifer I et al (2012) State of the art satellite and airborne marine oil spill remote sensing: application to the BP Deepwater Horizon oil spill. Remote Sens Environ 124:185–209. doi:10.1016/j.rse.2012.03.024

Liu Y et al (2011) Evolution of the loop current system during the Deepwater Horizon oil spill event as observed with drifters and satellites. In: Liu Y et al (eds) Monitoring and modeling the Deepwater Horizon oil spill: a record-breaking enterprise. AGU, Washington, pp 91–101

Mcginty S (2008) Fire in the night The Piper Alpha disaster. Pan Macmillan, London

Moller TH (1997) The NAKHODKA oil spill response – the technical adviser's perspective. Petroleum Association of Japan (PAJ) oil spill Ssmposium 97, 10–11 July 1997, Tokyo: MODIS: http://modis.gsfc.nasa.gov/

National Commission on the BP Deepwater Horizon Oil Spill and Offshore Drilling (2011) Deep water – the gulf oil disaster and the future of offshore drilling – report to the president. Jan 2011, pp 1–398
Noth EW et al (2011) Simulating oil droplet dispersal from the Deepwater Horizon spill with a Lagrangian approach. In: Liu Y et al (eds) Monitoring and modeling the Deepwater Horizon oil spill: a record-breaking enterprise. AGU, Washington, pp 217–226
OGJ editors (2015) Statoil's Gudrun condensate leak could have been major incident. Oil Gas J Online. 21 May 2015
Read C (2011) BP and the Macondo spill: the complete story. Palgrave Macmillan, New York
Reddy CM et al (2012) Composition and fate of gas and oil released to the water column during the Deepwater Horizon oil spill. Proc Natl Acad Sci U S A 109(50):20229–20234
Ryan JP et al (2011) A high-resolution survey of a deep hydrocarbon plume in the Gulf of Mexico during the 2010 Macondo blowout. In: Liu Y et al (eds) Monitoring and modeling the Deepwater Horizon oil spill: a record-breaking enterprise. AGU, Washington, pp 77–82
Ryerson TB et al (2012) Chemical data quantify Deepwater Horizon hydrocarbon flow rate and environmental distribution. Proc Natl Acad Sci U S A 109(50):20246–20253
Walker ND et al (2011) Impacts of loop current frontal cyclonic eddies and wind forcing on the 2010 Gulf of Mexico oil spill. In: Liu Y et al (eds) Monitoring and modeling the Deepwater Horizon oil spill: a record-breaking enterprise. AGU, Washington, pp 103–116
Weiberg RH et al (2011) Tracking subsurface oil in the aftermath of the Deepwater Horizon Well blowout. In: Liu Y et al (eds) Monitoring and modeling the Deepwater Horizon oil spill: a record-breaking enterprise. AGU, Washington, pp 205–216
White HK et al (2012) Impact of the Deepwater Horizon oil spill on a deepwater coral community in the Gulf of Mexico. Proc Natl Acad Sci U S A 109(50):20303–20308. doi:10.1073/pnas.1118029109
Wilson G et al (2013) Following the safe evacuation of all offshore personnel... the environmental response takes primacy: learning from the Elgin G4 Well major incident. SPE Offshore Europe Oil and Gas Conference and Exhibition, 3–6 Sept. Aberdeen. doi:http://dx.doi.org/10.2118/166590-MS

Chapter 3
Development and Operation of Underwater Robot for Autonomous Tracking and Monitoring of Subsea Plumes After Oil Spill and Gas Leak from Seabed and Analyses of Measured Data

Mahdi Choyekh, Naomi Kato, Yasuaki Yamaguchi, Ryan Dewantara, Hajime Chiba, Hidetaka Senga, Muneo Yoshie, Toshinari Tanaka, Eiichi Kobayashi, and Timothy Short

Abstract Oil spills produced by accidents from oil tankers and blowouts of oil and gas from offshore platforms cause tremendous damage to the environment as well as to marine and human life. To prevent oil and gas that are accidentally released from deep water from spreading and causing further damage to the environment over time, early detection and monitoring systems can be deployed to the area where underwater releases of the oil and gas first occurred. Monitoring systems can provide a rapid inspection of the area by detecting chemical substances and

---

M. Choyekh (✉) • N. Kato • Y. Yamaguchi • R. Dewantara • H. Senga
Naval Architecture and Ocean Engineering, Osaka University, Yamadaoka 2-1, Suita, Osaka 565-0871, Japan
e-mail: mahdi.choyekh@gmail.com; kato@naoe.eng.osaka-u.ac.jp; yamaguchi_yasuaki@naoe.eng.osaka-u.ac.jp; ryan.putra.dewantara@gmail.com; senga@naoe.eng.osaka-u.ac.jp

H. Chiba
Department of Merchant Marine, Toyama National College of Technology, 1-2 Ebieneriya, Imizu, Toyama 933-0293, Japan
e-mail: chiba@nc-toyama.ac.jp

M. Yoshie • T. Tanaka
New technology Development Field, Port and Airport Research Institute, 3-1-1 Nagase, Yokosuka, Kanagawa 239-0826, Japan
e-mail: yoshie@pari.go.jp; tanaka_t@pari.go.jp

E. Kobayashi
Graduate School of Maritime Sciences, Kobe University, 5-1-1 Fukaeminami, Higashinada-ku, Kobe, Hyogo 658-0022, Japan
e-mail: kobayasi@maritime.kobe-u.ac.jp

T. Short
Marine and Space Sensing, SRI International, 450 Eighth Avenue SE, St. Petersburg, FL 33701, USA
e-mail: timothy.short@sri.com

© Springer Japan 2017

N. Kato (ed.), *Applications to Marine Disaster Prevention*,
DOI 10.1007/978-4-431-55991-7_3

collecting oceanographic data necessary for enhancing the accuracy of simulation of behavior of oil and gas. An autonomous underwater vehicle (AUV) called the spilled oil and gas tracking autonomous buoy system (SOTAB-I) has been developed to perform on-site measurements of oceanographic data as well as dissolved chemical substances using underwater mass spectrometry. In this chapter, the outlines of SOTAB-I and a description of its hardware and software are presented. The operating modes and guidance and control of the robot are detailed. The experimental results obtained during the early deployments of SOTAB-I in the shallow water of the Gulf of Mexico in the USA demonstrated the ability of SOTAB-I to collect substances' dissolutions in seawater such as hydrocarbons. Deepwater experiments were conducted in Toyama Bay in Japan and enabled demonstration of the ability of SOTAB-I to establish the vertical water column distribution of oceanographic data, such as temperature, salinity, and density. In addition, a high-resolution profile of water currents was obtainable.

Keywords AUV • Survey • Chemical substances • Oceanographic data

3.1 Introduction

The world economy depends to a large extent on the use of energy. To meet the increasing need for energy, both in industry and daily life, petroleum activities, such as drilling and shipping, are on the rise. Thus, additional attention is required to avoid accidents that can happen due to such activities. Oil spills and blowouts of oil and gas from the seabed cause serious damage to the environment as well as to the economy, not to mention the damage to marine and human life. For the case in which methane gas is blown out from a seabed, it is partly dissolved in seawater and then partly consumed by methanotrophs (Kessler et al. 2011), which leads to the creation of local hypoxia zones caused by oxygen depletion (Shaffer et al. 2009). The rest of the gas is released to the atmosphere, contributing to global warming, as methane is a highly potent greenhouse gas (Solomon et al. 2009). Recently, several oil spill accidents have happened. Deepwater Horizon in the Gulf of Mexico in 2010 and the Elgin gas platform in the North Sea in 2012 are examples of these accidents. To prevent oil and gas spills from spreading and causing further damage to the environment over time, early detection and monitoring systems can be deployed around the offshore oil and gas production system. In addition, oceanographic data should be collected to comprehend the environmental changes around the accident. Based on the collected information, oil and gas drifting simulations must be performed to predict where the spilled oil will wash ashore and to adequately deploy oil recovery machines before this occurs.

A wide variety of methods that deal with underwater oil spills exists, and each presents strengths and weaknesses according to the circumstances and the purposes for which it is deployed. For substances' dissolution measurement, among the most commonly used techniques is the extraction of discrete samples for subsequent analyses (Joye et al. 2011). However, this method has limited temporal and spatial resolution. Additionally, it requires much effort and is time consuming. Furthermore, a risk is that the characteristics of the original collected samples could change during the collecting and handling processes. Other techniques are utilized to track a particular substance, such as oxygen, methane, or carbon dioxide. These methods can provide continuous information regarding the dissolution of substances, but only for a particular and limited variety of substances. The spilled oil and gas tracking autonomous buoy system (SOTAB-I) integrates an underwater mass spectrometer (UMS) that overcomes the previously mentioned weaknesses. The UMS enables real-time on-site measurements. It is distinguished by its good flexibility and sensitivity as well as its high reliability. It can detect multiple substances' dissolutions simultaneously (Short et al. 2006).

The challenge in water surveying is not only to detect oil and substances dissolved in seawater but also to obtain other related oceanographic data, as many research programs have demonstrated that temperature (Servio and Englezons 2002), pressure (Handa 1990), and salinity (Yang and Xu 2007) are critical factors that considerably affect the formation and dissociation of gas hydrate. In addition, measurement of underwater currents is important for detecting and tracking dissolved gases and for predicting the evolution of the blowout gas in simulation models. Few existing compact systems are able to conduct a complete survey that can measure salinity, temperature, and depth as well as underwater currents and dissolved gases simultaneously. In Deep Spill experiments in Norway, for example, substances' dissolutions were collected using a rosette with sampling flasks towed from a ship. Underwater current data were collected from two separate acoustic Doppler current profiler (ADCP) devices, one on board a second ship and the other connected to an acoustic transmitter and moored on the seabed (Johansen et al. 2003). Using such a technique is helpful to obtain a full survey of the area. However, it requires good synchronization while operating because sensors are mounted in different places. In addition, it requires further resources for deployment, which increases the cost of the survey. The SOTAB-I combines necessary sensors for a full and complete real-time and on-site survey by integrating a UMS, an acoustic Doppler current profiler (ADCP), a conductivity-temperature-depth sensor (CTD), and a camera.

Another challenge related to underwater surveying efforts is the range of the survey. For example, for water profiling, few systems that can provide a deepwater profile exist. Existing acoustic water profiling sensors based on ADCP can barely perform measurements of water current distribution beyond the limit of 1000 m depth from the sea surface. Furthermore, in most ADCP devices, resolution of the water layer decreases when higher ranges are applied. Some investigations could lead to a longer range by using multiple ADCP devices. For instance, a solution that implements a dual-meter system based on a surface- and a bottom-mounted ADCP was suggested (Vogel et al. 2001). The system could measure the water currents' profile up to 1200 m. However, the system lacks the flexibility needed for prompt intervention missions. The SOTAB-I has the capability to perform a high-resolution survey with extended range from sea surface up to a water depth of 2000 m.

Due to their compactness, the use of AUVs for full-water surveying is being adopted increasingly (Jakuba et al. 2011; Harvey et al. 2012). Among the existing types of underwater robots used to autonomously monitor marine environments in 3D space from sea surface to seabed over the long term is the Argo Float (Roemmich et al. 2009) that floats vertically and repeats descending and ascending in the vertical direction by using a buoyancy control device. However, it does not have a function of active movement in the horizontal direction. Another method is the underwater glider (Eriksen et al. 2001), which has a streamlined body with fixed wings. It can descend and ascend also by using a buoyancy control device, while it moves in the horizontal plane like a glider for long distance. However, the ratio of vertical movement distance to horizontal movement distance is small. SOTAB-I was designed to move not only in the vertical direction by a buoyancy control device but also in the horizontal direction by two pairs of rotational fins.

The development of a new type of AUV requires an evaluation process from two aspects. One is the guidance and control of the vehicle and the other is the data sampling. In the first part of this chapter, descriptions of the underwater robot SOTAB-I and its hardware and software are presented. In the second part, the maneuverability of the SOTAB-I is described. The third part shows the survey results obtained in shallow water in the Gulf of Mexico in the USA and in deep water in Toyama Bay in Japan.

3.2 SOTAB-I Overview

3.2.1 Outlines of SOTAB-I

The SOTAB-I is 2.5 m long and weighs 325 kg. It can be submerged in water as deep as 2000 m. It is able to descend and ascend by adjusting its buoyancy using a buoyancy control device while changing its orientation through two pairs of movable wings. The SOTAB-I can also move in horizontal and vertical directions using two pairs of horizontal and vertical thrusters. A visual overview of SOTAB-I is illustrated in Fig. 3.1, and its main characteristics are summarized in Table 3.1. The arrangement of devices and sensors installed on SOTAB-I is shown in Fig. 3.2.

When the robot floats on the sea surface, a wireless local area network (WLAN) and an iridium satellite communication transceiver module are used for data transmission. When the robot is underwater, the user on the mother ship and the SOTAB-I can communicate through the acoustic modem.

The robot tracking on the sea surface is ensured by a global positioning system (GPS) receiver that serves to determine the robot's absolute position. In the case where the robot is submerged, tracking is ensured by the ultrashort baseline (USBL) system. The vertical position of the robot in the water column is given by depth data from the CTD sensor. When the robot is within the bottom tracking altitude from the seabed, the Doppler velocity logger (DVL) is able to measure robot's velocities. The robot motion and orientation are given by the compass and the inertial measurement unit (IMU). An ADCP is employed to measure the magnitude and orientation of water current layers. SOTAB-I is also fitted with a UMS to

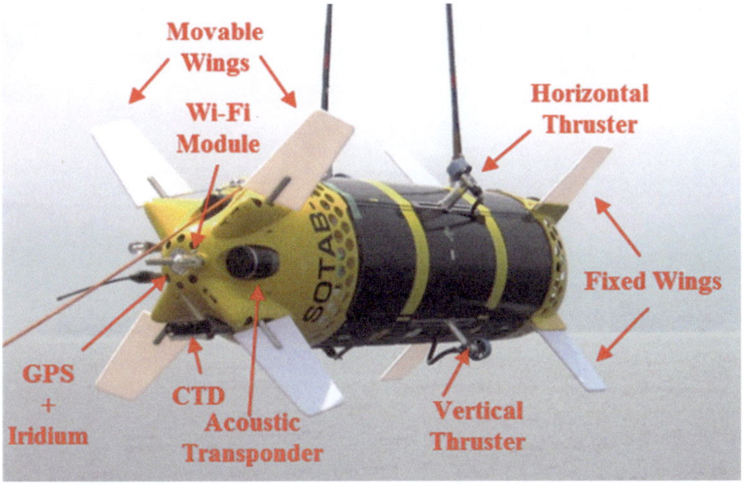

Fig. 3.1 SOTAB-I robot

Table 3.1 Principal particulars of SOTAB-I

Total length [mm]		2503
Diameter [mm]		667
Weight in air [kg]		311.7
Weight in water [kg]		±3.8
Movable wing	Chord [mm]	200
	Span [mm]	400
Fixed wing	Chord [mm]	200
	Span [mm]	400

determine the characteristics and physical properties of the dissolved gas and oil. To obtain a visual representation of blowouts of plumes of gas on the seabed, the robot is equipped with a camera.

3.2.2 Hardware Description

SOTAB-I devices can be classified based on their function, as shown in Fig. 3.3. Following are further details about SOTAB-I hardware.

3.2.2.1 Power Supply

SOTAB-I is powered by 32 packs of lithium-ion batteries divided into pairs and mounted in a serial arrangement. The 16 pairs are then mounted in parallel, which gives a voltage of 28.8 V because every single pack voltage is 14.4 V.

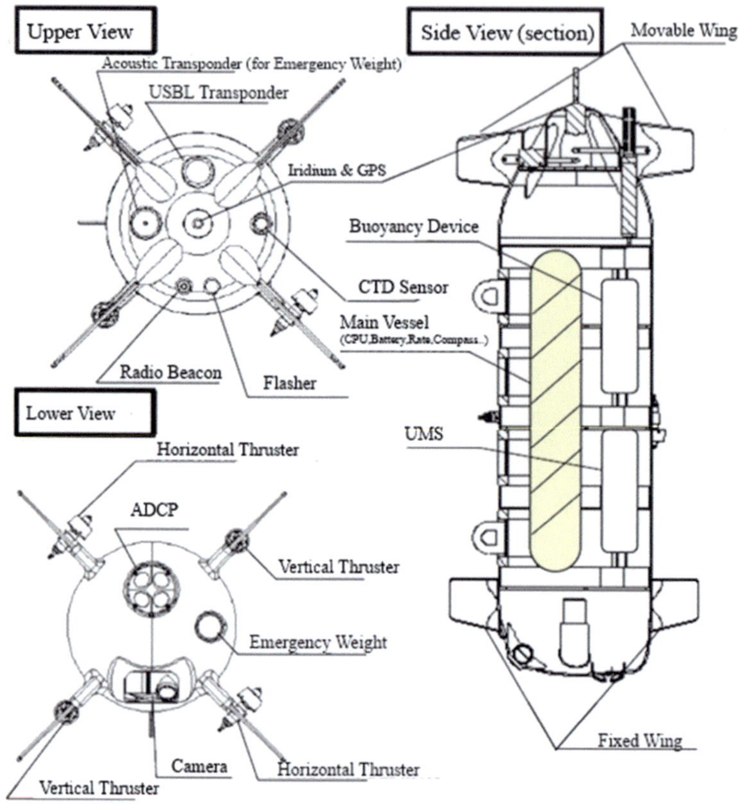

Fig. 3.2 Arrangement of devices and sensors installed on SOTAB-I

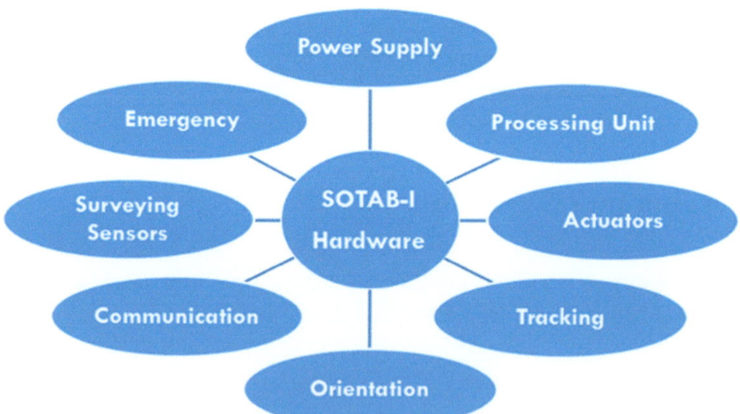

Fig. 3.3 Hardware classification

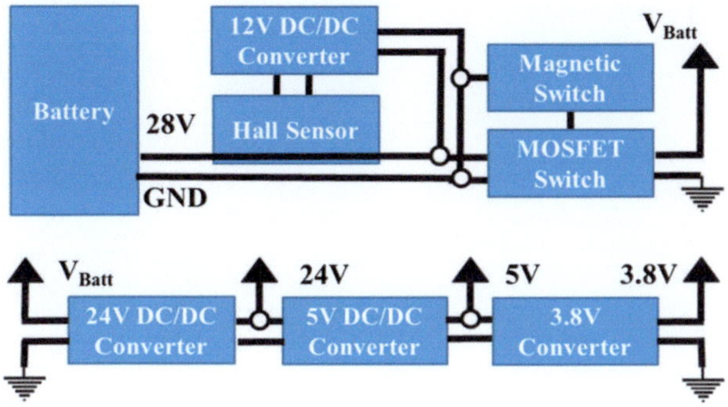

Fig. 3.4 Power supply conversion

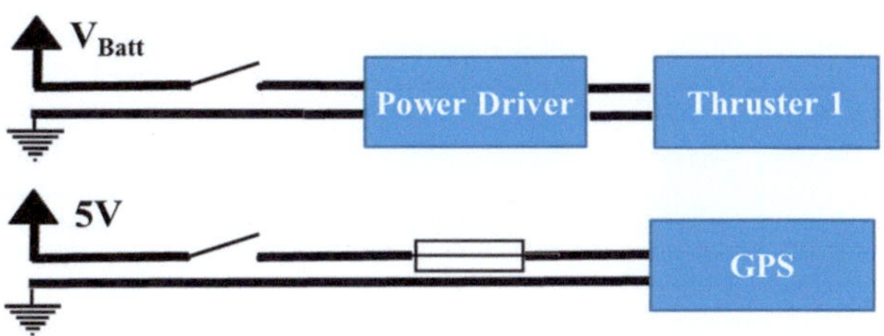

Fig. 3.5 Power supply of actuators and sensors

The total capacity of the batteries is equal to 4608 W•h. Table 3.11 presents the characteristics of the batteries used. As shown in Fig. 3.4, the power is activated via an external magnetic switch that controls the gate of the power MOSFET (metal-oxide-semiconductor field-effect transistor), used here as a power switch. DC/DC converters are employed to convert the battery voltage output to the different levels of power supply required for the operation of the internal devices. A hall sensor is used to measure the current. Two power switch control boards serve to control the power supply of internal devices. They also include necessary fuses used for protection against surcharges. Power activation of each sensor/actuator is controlled by the processing unit board I/O port. Control boards and power boards are optically separated. Figure 3.5 shows a typical electronic circuit used for powering the GPS sensor. The same concept is used for all other sensors, with the only difference in the input voltage. An example of a circuit used to power the thruster 1 is also shown in Fig. 3.5. Similarly, all actuators such as thrusters, wings, and buoyancy device have the same power supplying concept. Table 3.2 shows the power consumption of sensors installed on SOTAB-I.

3.2.2.2 Processing and Control Unit

The processing unit of the SOTAB-I is made up of two processors. The volatile memory capacity is 1 GB. The characteristics of the processing unit are detailed in Table 3.12. The board has an Ethernet port connected to a Wi-Fi router that serves to connect to a remote computer through a wireless local area network (LAN). The

Table 3.2 Power consumption of sensors

Sensor	Power consumption
CTD	3.4 W (typical)
UMS	60–80 W
DVL	3 W (typical)
USBL	100 W (during transmission)
Iridium	1 W
GPS	0.4 W (typical)
IMU	0.22 W (maximum)
LAN	2.3 W (maximum)
CPU	16.27 W (typical), 25.3 W (maximum)

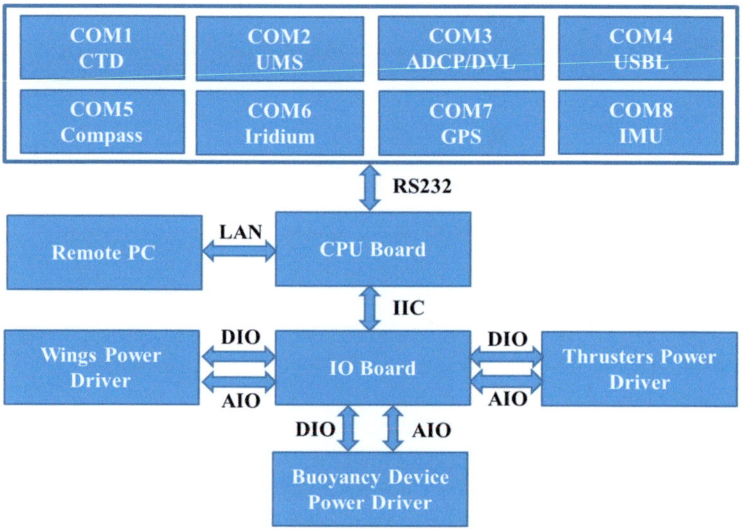

Fig. 3.6 Computer communication interfaces

processing unit is also fitted with several interfaces such as USB, RS-232, and inter-integrated circuit (I^2C). Three of the USB ports are used to connect external flash memories. The SOTAB-I computer is composed of three storage drives. The C drive is used for the operating system, the E drive for the programs' setup files, and the D drive for logging and program storage.

All sensors are directly interfaced through their respective RS-232 serial ports. Internally, the CPU board connects to an input/output board (IO board) through the inter-integrated circuit (I^2C) serial interface. The IO board is composed of circuits that are able to interface with 32 digital input outputs (DIO), 8 analog outputs using digital-to-analog converters (DAC), and 8 analog inputs (AI) using analog-to-digital converters (ADC). The IO board ensures isolation between the CPU board and the power drivers. CPU board interfaces are summarized in Fig. 3.6.

3.2.2.3 Actuators

Buoyancy Device

In the buoyancy control device, an oil hydraulic pump injects and extracts oil between the external oil bladder and the internal oil reservoir. A motor valve serves to automate opening and closing cycles, and a brake is used to lock the pump (Fig. 3.7). The flow rate during the injection of oil into the bladder at the external pressure of 20 MPa is 243 mL/min, and during the extraction of oil from the bladder at the same external pressure condition, it is 349 mL/min. In total, six digital inputs are employed to control the buoyancy device. A digital input serves to control the power supply relay. One input is used to open the valve and another to close it. To control the motor pump actuator, one input is used to run/disable it and another one specifies the rotation direction. One more serves to activate/deactivate the brake. The feedback is provided by two digital outputs that report the valve position and one analog output that provides the oil level.

Figure 3.8 depicts the relationship between the pressure in the reservoir and the volume of the oil in the reservoir. The "pressure in the reservoir" is the pressure applied by the hydraulic pump to the oil reservoir cylinder. The maximum allowable drain pressure of the pump is 0.03 MPa. The oil reservoir is fitted with a linear potentiometer whose analog output voltage is an image of the piston displacement under the pump drain pressure. The voltage is proportional to the oil volume of the reservoir, which leads to the variation of robot buoyancy, as shown in Table 3.3. When the oil room is full, the output voltage is 0.3892 V. When the oil room is empty, the output voltage is 1.6208 V. Because the reservoir has a cylindrical shape and the surface of the base is constant, the output voltage of the potentiometer can be used to determine the volume of the oil in the reservoir, which explains the linear volume/output voltage relationship. The mass change can be obtained after

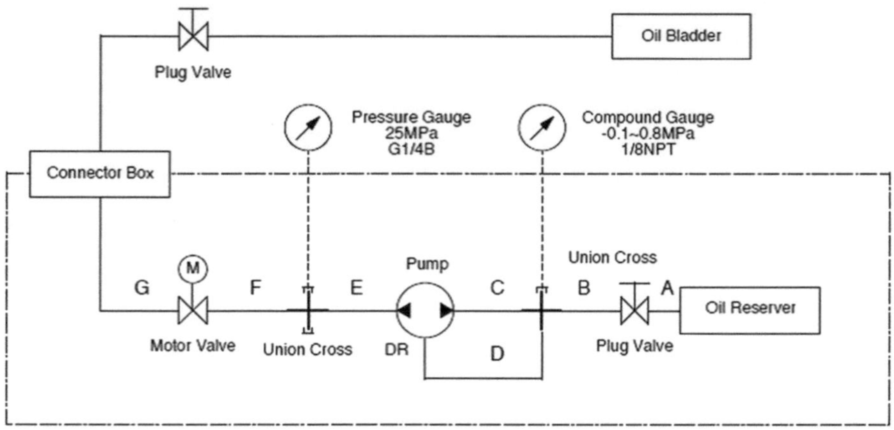

Fig. 3.7 Buoyancy device

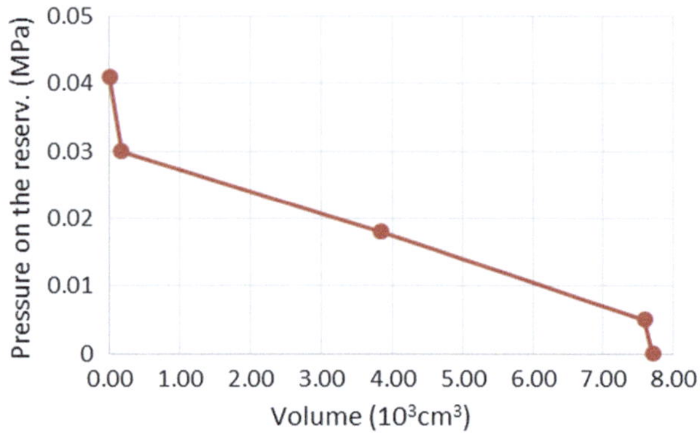

Fig. 3.8 Relationship between the pressure and the volume in the oil reservoir

Table 3.3 Variation of robot buoyancy against voltage on motor

Mass change (seawater) (g)	Volume change (cm^3)	Potentiometer output voltage (V)	Pressure (MPa)
7,884.8	7,700.0	1.621	0
7,746.6	7,565.0	1.599	0
7,763.2	7,581.2	1.602	0.005
3,925.7	3,833.7	1.012	0.018
171.8	167.8	0.443	0.030
0	0	0.389	0.041

multiplying the volume change with the density of seawater, which is typically equal to 1024 kg/m^3.

Wings

The robot is equipped with two pairs of rotational wings situated at the top side of the robot and four fixed wings attached at the bottom side. Each pair of movable wings is controlled around a rotational axis with the same direction and angle by a servomotor, and the rotational axis of one pair of movable wings is perpendicular to that of the other. DC motor drives connected to a gearhead with a 53:1 ratio enable each pair of wings to move from −90 to 90°. A rotational potentiometer with a resolution better than 0.1° provides information about the current orientation of the wings. A separate microcontroller control board integrating a proportional-integral-derivative (PID) controller program is used to control the wings through a serial port. The typical power consumption of the wing drive is 7 W.

Thrusters

Thrusters are controlled by an analog input varying from 0 to ±5 V. A relay is used for turning the power on and off. A tachometer measures the rotational speed and converts it to an analog output. Each thruster has a maximum thrust force equal to 35.28 N forward and 26.46 N backward. Table 3.12 gives more details about thruster specifications.

3.2.2.4 Tracking

The robot absolute position is needed for several purposes, notably for robot guidance and water survey. Robot control relies on underwater position to determine the direction and distance in which the robot should go. The concentration of oil substances will be assigned to their absolute position. Then it is possible to know the distribution of oil in the area and use that to develop a strategy to track the oil slick or find the blowout gas source. SOTAB-I is fitted with a GPS receiver that can output the robot's absolute position when it is on the sea surface. DGPS accuracy is better than 3 m with 95 % confidence. The characteristics of the GPS receiver are described in Table 3.14. The high-frequency waves of the global positioning system (GPS) have very low penetrability into water. Instead, acoustic-based positioning systems (APS) are used widely for underwater tracking. Several APS techniques can be used (Vickery 1998). SOTAB-I integrates an ultrashort baseline (USBL) system for underwater positioning. It has the advantage of being very easy to deploy because it only requires a single transceiver on the ship side and one transponder installed on the robot side. In addition, USBL systems have an acceptable range and accuracy. The USBL system deployed with the SOTAB-I can track up to eight targets at a range of up to 5000 m with ship noise. Its accuracy is equal to 0.15° for the bearing angle accuracy and 0.3 m for the slant range. On board the ship, a GPS pole is deployed with a vertical reference unit (VRU) system to determine the absolute position of the ship in addition to the tilting of the acoustic transceiver. The GPS compass has a DGPS horizontal accuracy better than 0.6 m with 95 % confidence. Heading accuracy is better than 0.15° rms. Further details about the GPS compass are described in Table 3.15. The relative underwater position of SOTAB-I to the mother ship is combined with the ship absolute position to determine the absolute position of the robot in the Earth-fixed coordinates. USBL determines the relative position by calculating the range and the angle between the transceiver and the transponder. At first, an acoustic pulse is transmitted by the transceiver. When the robot's transponder detects the pulse, the transponder replies by sending another pulse. The time difference between the transmission of the transceiver pulse and the arrival of the transponder pulse is converted to range, while the phase difference within the transceiver's transducer array determines the angle.

3.2.2.5 Orientation

SOTAB-I is fitted with a tilt-compensated compass module that provides attitude information. The accuracy of the compass module is 0.3 °rms for the heading angle and 0.2 °rms for the tilting angles. The maximum sampling frequency is 20 Hz, and its typical power consumption at that rate is 0.1 W. The module is accompanied with hard and soft iron calibration algorithms that contribute to the improvement of the reliability and consistency of heading measurement. Detailed features of the IMU sensor are detailed in Table 3.20.

SOTAB-I is also equipped with an inertial measurement unit (IMU) and attitude heading reference system (AHRS) that combine three-axis accelerometers, three-axis gyros, three-axis magnetic sensors, a barometric pressure sensor, and a 32-bit built-in microcontroller. The IMU microcontroller runs a quaternion-based extended Kalman filter (EKF), which provides estimates of the attitude of the sensor as well as the real-time gyro biases. It can output acceleration, angular rate, and magnetic measurements along the body coordinate frame as well as in the north–east–down (NED) coordinate frame. The device can also measure the temperature and the barometric pressure. The sensor has already been subject to a factory calibration. Additionally, it provides the user with the ability to apply a separate user calibration to remove additional bias, scale factor, and axis misalignments. The sensor includes a separate EKF that provides real-time estimation of the local magnetic hard and soft iron distortions. The maximum power consumption of the module is 0.22 W. Data measurement update rate, including Kalman filter data processing, can go as high as 300 Hz. The communication with the IMU sensor is performed through an RS-232 serial port, but it is also possible using the SPI interface. The system was configured to output binary message data, which contains the maximum amount of information.

3.2.2.6 Communication

The communication between the SOTAB-I and the mother ship depends on the distance between them and whether the robot is on the sea surface or underwater (Fig. 3.9). When the SOTAB-I is on the sea surface and within a 200 m range of the mother ship, it is possible to remotely connect to the robot computer and take full control of the robot. The UDP port is used to exchange data, such as the state of the joystick control button, between the ship GUI and the robot GUI. SOTAB-I is also fitted with an Iridium satellite communication transceiver module. It has global coverage and can send and receive messages within 6–22 s. Transmission speed is up to 340 bytes per message, and reception speed is up to 270 bytes per message. When the robot is underwater, real-time communication with the ship using radio waves becomes impossible, and acoustic communication is used instead. In the current configuration, the acoustic communication speed is equal to 1320 bytes per 8 s for uplink and 80 bytes per 8 s for downlink. To be able to establish acoustic

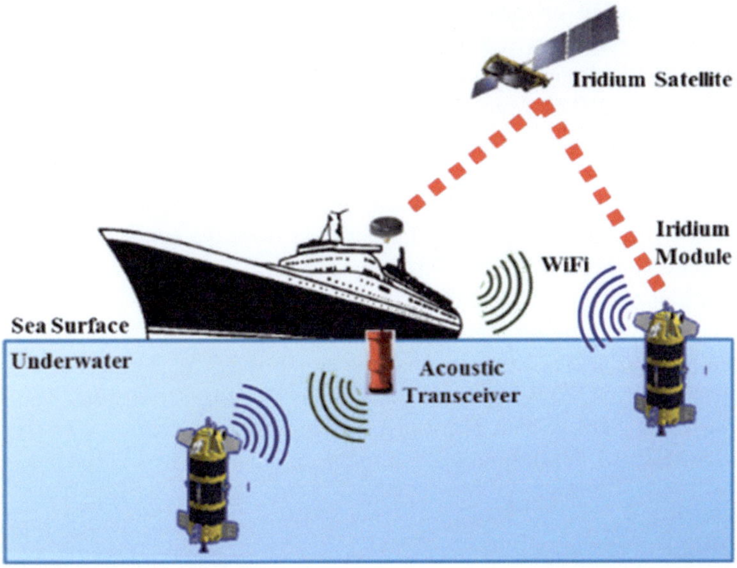

Fig. 3.9 Communication

communication, it is important that the robot be within the operating beam width of the transceiver, which is equal to 120°.

3.2.2.7 Surveying Sensors

CTD

The CTD sensor can measure the temperature, conductivity, and pressure of seawater. Based on those measurements, it is possible to calculate additional oceanographic data such as salinity and density as well as sound speed. The vertical position of the robot in the water column can also be obtained based on the pressure value. The CTD sensor employed has a sampling frequency up to 16 Hz, enabling a very high spatial resolution. It has a high accuracy and a typical power consumption of 3.4 W. A serial RS-232 communication port interfaces with the sensor. The characteristics of the sensor are described in Table 3.16.

ADCP/DVL

An ADCP/DVL device was used for water profiling and bottom tracking. It integrates heading and attitude sensors that are necessary for coordinate transformation. The accuracy is within ±2° for the compass and ± 0.5° for the attitude.

An integrated thermistor measures water temperature and improves the accuracy of calculation of sound speed as well as enhances the accuracy of the acoustic measurements. The device is mounted looking downward at the bottom of the robot, as shown in Fig. 3.2. The device has four piston transducers with standard acoustic frequency equal to 1228.8 kHz, which enables high-resolution measurements of water currents up to 13 m range at 0.5 m layer resolution. The number of layers is selectable, and the layer thickness can be customized from 0.25 to 5 m. When the robot is within the bottom tracking altitude from the seabed (23 m when the device is powered at 24 V), the Doppler Velocity Log (DVL) can determine the robot velocities as well as the altitude from the seabed. The average power drawn by the ADCP/DVL pinging at its maximum rate at maximum altitude is 3 W, with a constant background power consumption equal to 2.2 W. The robot's processor connects to the ADCP/DVL device through an RS-232 serial port. The selected output format is PD0, which is a binary format that provides the most possible information. Table 3.17 summarizes the main characteristics of SOTAB-I ADCP.

UMS

SOTAB-I is equipped with an underwater mass spectrometer (UMS). The UMS instrument used for the SOTAB-1 deployments contained a 200 amu linear quadrupole mass analyzer (E3000, Inficon, Inc., Syracuse, New York). Table 3.18 provides the specifications of SRI International membrane introduction mass spectrometry (MIMS) system.

Introduction of analytes into the mass spectrometer occurs through a hydrophobic and nonporous high-pressure polydimethyl siloxane (PDMS) membrane introduction system, pressure tested to a depth of 2,000 m. Water samples are placed in contact with the semipermeable membrane, usually at a constant flow rate. The transport of dissolved gases and relatively nonpolar volatile organic compounds (VOCs) through these membranes is compound specific and temperature dependent, but typically requires that the solute dissolves into the membrane, diffuses through it, and finally evaporates into the mass spectrometer. Once in the mass spectrometer vacuum chamber, the neutral gas-phase analytes are (1) ionized by electron impact, (2) sorted by their mass-to-charge (m/z) ratios (typically $z = 1$), and (3) detected to create a mass spectrum. The membrane interface used in this system provides parts-per-billion level detection of many VOCs and subparts-per-million detection limits for many dissolved light stable gases.

The membrane probe assembly consists of a hollow fiber PDMS membrane stretched and mounted on a sintered Hastelloy C rod. One end of the supported membrane is capped with a polyetheretherketone (PEEK) rod; the other end is connected to the vacuum chamber via stainless steel tubing. The membrane assembly is inserted into a steel heater block that houses a thermocouple and heater cartridges for controlling sample and membrane temperature (± 0.1 °C). A magnetic piston pump draws ambient water into the sample tubing, through the membrane probe assembly, and back to the environment.

3.2.2.8 Emergency

Since the communication with the robot may be lost for various reasons, such as a software crash or a hardware problem, commanding the robot may be no longer possible. In this particular case, a second method based on the use of an acoustic weight cutoff unit, which is completely independent from the circuit of the system, can be utilized for emergency surfacing. An additional radio beacon and a blinker flasher light are installed in the robot to assist in finding it in such circumstances.

3.2.3 Software Description

The SOTAB-I performs its surveying tasks autonomously. The operating mode is selected by the user on board the mother ship at the beginning of the operation by means of a graphical user interface (GUI). Orders are transmitted to the underwater robot through the acoustic modem. On the robot side, oceanographic and dissolved substance data collected by the SOTAB-I are sent to the mother ship in real time through an acoustic modem. On the mother ship side, the GUI receives the data collected from SOTAB-I, stores them, and displays the most essential data to the user (Fig. 3.10). Data related to the spilled plume and underwater currents are processed not only for guidance and control of the SOTAB-I but also for the simulation and the prediction of plume behavior contributing to the decision-making process for the best deployment strategy of collecting and cleaning machines.

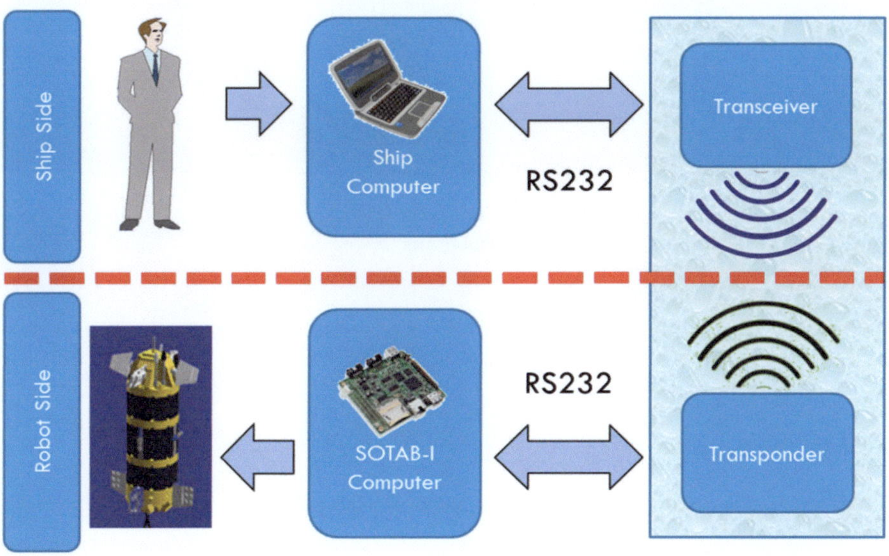

Fig. 3.10 Diagram of operation during at-sea experiments

3.2.3.1 Ship Computer

On board the ship, a portable computer running on Windows 7 operating system is employed. The integrated development environment (IDE) is Borland C++6. The GUI and the TrackLink will be executed. TrackLink is the software that comes with the USBL product.

TrackLink is used to determine the robot position and to manage the underwater communication. It connects to the external devices through an RS-232 serial interface (Fig. 3.11). TrackLink output data follows the LQF format. The data include information about the ship position and heading as well as the robot position and its depth. TrackLink and the ship GUI exchange information through virtual RS-232 serial ports. The ship GUI sends downlink data to the robot via the TrackLink software. From the GUI side, the Iridium module and the hydrophone are interfaced through a serial RS-232 interface. A joystick is interfaced through the USB port. The GUI (Fig. 3.12) has several roles that can be classified to three categories: display, guidance, and communication. The GUI shows the most important information about the robot status and its environment. It displays the robot position, orientation, and speed in addition to its actuators' status. Environmental data such as oceanographic data and substances' dissolutions are also displayed. Additionally, the GUI enables the selection of the operating mode and the manual control of the actuators. It gives the possibility to send a software emergency ascending order. The GUI combines interfaces for managing the Iridium satellite communication and the acoustic communication.

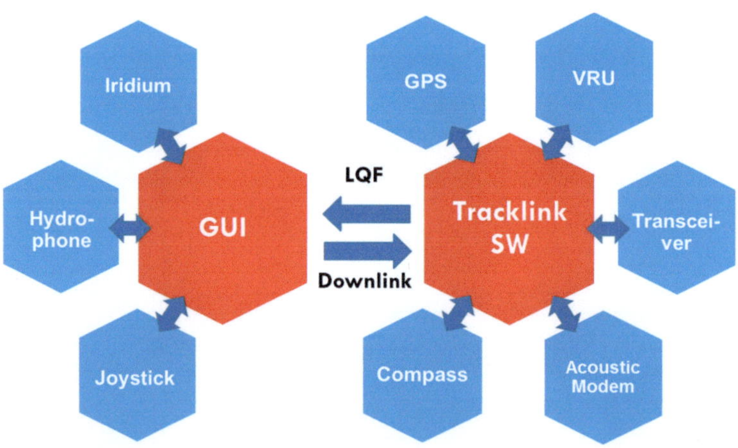

Fig. 3.11 Ship computer

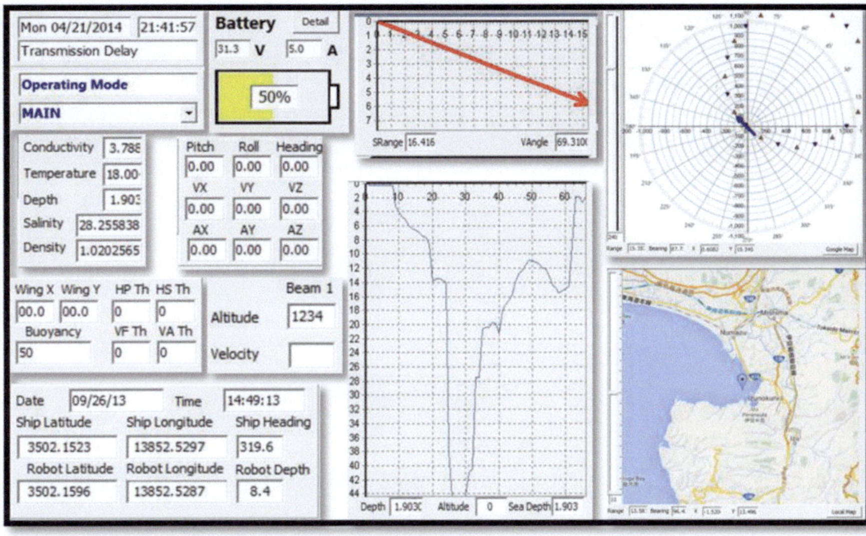

Fig. 3.12 Overview of SOTAB-I GUI

3.2.3.2 SOTAB-I Computer

The SOTAB-I program is running on the Windows XP embedded operating system. Borland C++ Builder 6.0 was selected as the integrated development environment (IDE) to develop the GUI. At the beginning of the operation, the magnet that is put on the magnetic switch is removed and the system is powered on. After the operating system is started, the executable program of the GUI is launched automatically. The average CPU usage when the program is fully executing is 3 %.

The robot processor connects to the sensors through RS-232 serial ports. Because a physical serial port can only be used by a single application at one time, a virtual serial splitter was employed to duplicate the serial data input (Fig. 3.13). One of the duplicated inputs is directed to a serial logger software to save raw data in a file for ulterior detailed analysis. Another one is sent to the main program for real-time processing. All sensors' data are saved by the GUI every 1 s in their corresponding files with their associated time, which makes it easier to synchronize at post-processing. A third clone of the serial input is directed to a third-party software that is specific to each sensor.

The GUI software was organized into a hierarchical order as shown in Fig. 3.14. In this multilayered architecture, responsibilities are distributed. Drivers are the software libraries that directly interface with the hardware and control it. They initialize the device and manage it. The driver layer contains drivers for peripherals such as DAC, ADC, I^2C interface, RS-232 interface, etc. The real-time operating system (RTOS) can run multiple tasks simultaneously. The task can exchange information through the global variables (Fig. 3.15). All tasks are executed peri-

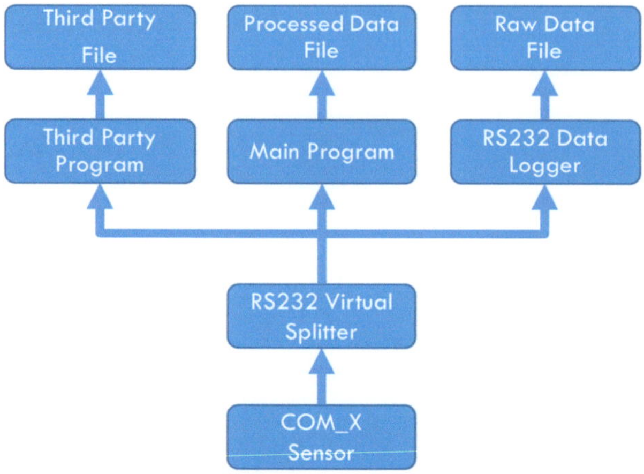

Fig. 3.13 Serial data acquisition

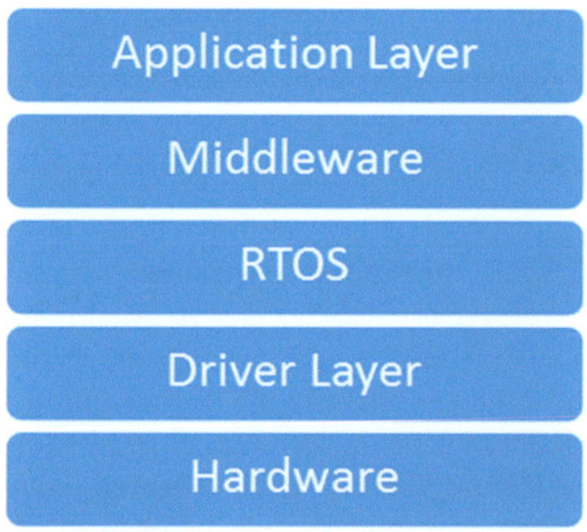

Fig. 3.14 Software architecture of SOTAB-I GUI

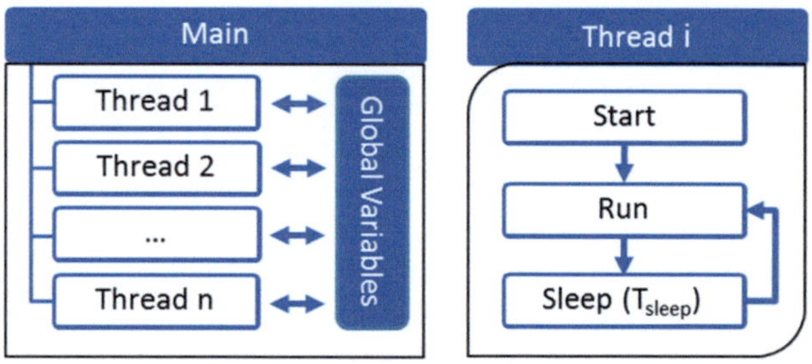

Fig. 3.15 Application layer, general structure of thread execution

odically. A sleep command is used to determine the frequency of the execution of the thread. At the beginning of the execution of the program, every task is initialized. The current thread runs sequentially through the instruction till the end. Then, the execution of the current thread is stopped until at least Tsleep has elapsed. The middleware layer provides services to the upper application to facilitate the communication, input/output, and data management. It is responsible for managing serial communication and LAN and ensures process-to-process communication. The application layer contains all personalized programs of the robot.

The program is composed of several files that can be classified into three categories: the header files that include all the function definitions and predefined variables, the source files that include the program source of function, and the configuration files, which are the only files that need to be modified by the user.

3.2.3.3 Acoustic Communication

The acoustic data exchange format between the ship and the SOTAB-I is defined as shown in Fig. 3.16. Downlink data are used to command the SOTAB-I. In addition, it is used for sending the robot's absolute underwater position. For that reason, downlink data are sent continuously. At the beginning of the experiments, SOTAB-I and the ship computer's times are synchronized. Using the time stamp that is associated with each downlink data, it is possible to determine the shifting between the time in which the robot position was determined on the ship and the time when it was received by the robot. Uplink data include sensor data and actuator states to keep the user informed about the robot situation and the changes in its environment. Downlink data are also feedback on the uplink data to verify the authenticity of

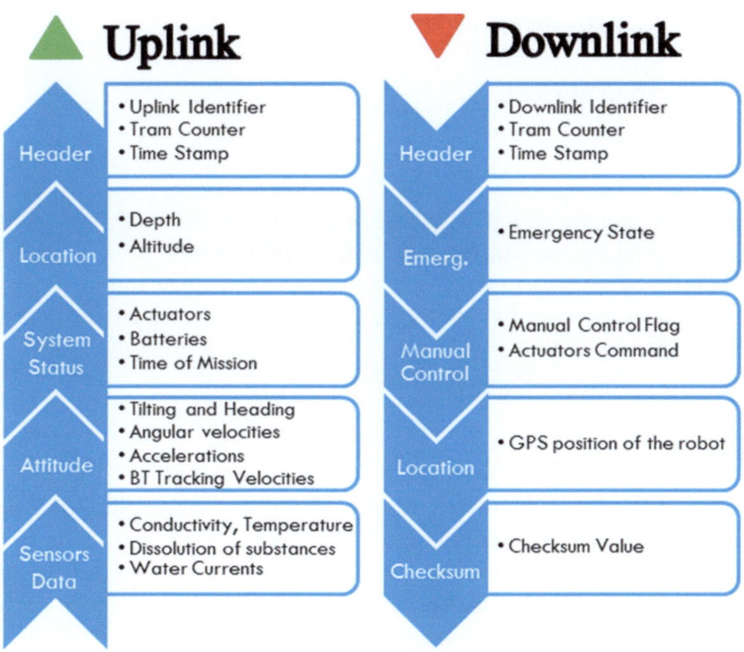

Fig. 3.16 Acoustic communication data frame

the communication and whether noise occurred during data submission. The ASCII format was selected for coding data. This gives the possibility to read and send the data simply by using a serial terminal. Additionally, data can be interpreted without a specific parser. The binary format gives the advantage of sending a bigger amount of data. However, uplink and downlink data exchange rates are enough to send all important data under the text format.

During the experiments, it was noticed that additional bytes are sometimes injected into the downlink data. Since the downlink data are used to control the actuators of the robot, noise injected into the data may lead to a misinterpretation of the orders and cause serious damage. For instance, if the robot receives an order to move the vertical thruster in the backward direction, the robot will start diving very fast to the seabed, and it may be impossible to recover it. To enhance the immunity of the system against mistaken orders caused by noise, a checksum algorithm was used for downlink data. The integer addition checksum was employed. It is a simple algorithm that reduces considerably the probability of undetected errors. It can detect all single bit errors and all error bursts of length 16 bits or less. The percentage of undetected 2-bit errors over the total number of 2-bit errors is less than 3 % for a message length equal to 80 bytes (Maxino and Koopman 2009).

3.3 SOTAB-I Guidance and Control

3.3.1 General Description

3.3.1.1 Operating Modes

Manual Mode

In this mode, the control of the robot will be performed manually through the GUI. Commands can range from simple orders to following a whole control scenario. This mode primarily enables us to test the robot's basic functions and to verify that all sensors and actuators as well as tracking and data transmission devices are working correctly.

Survey Mode

SOTAB-I has three main surveying modes. At the first stage, SOTAB-I performs the water column survey by adjusting its buoyancy. The rough mode is used to collect rough data on physical and chemical characteristics of plumes by repeating descending and ascending on an imaginary circular cylinder centered at the blowout position of oil and gas through the variation of buoyancy and movable wings' angles. Finally, in case the UMS detects a high concentration of any particular substance, a precise guidance mode will be conducted to track and survey its detailed characteristics by repeating descending and ascending within the plume.

Photograph Mode

This mode enables us to have a large visual overview of the area around the blowout position of oil and gas by taking pictures of the seabed and making image mosaicking. SOTAB-I moves laterally using horizontal thrusters along diagonal lines of a polygon with a radius of 5 m centered on the blowout position of oil and gas.

3.3.1.2 Operating Zones

Based on the robot altitude from the seabed, the water column is divided into three zones, as illustrated in Fig. 3.17:

Normal Operation Zone

In this zone, the robot performs surveying in operational modes as well as the photograph mode. When the user on the mother ship decides to launch a specific operational mode, the GUI reads its associated ID and includes it in the downlink data that will be sent through the acoustic modem. The robot receives the data, identifies the requested mode, and executes it. At the end of the execution, the robot waits for the next downlink order.

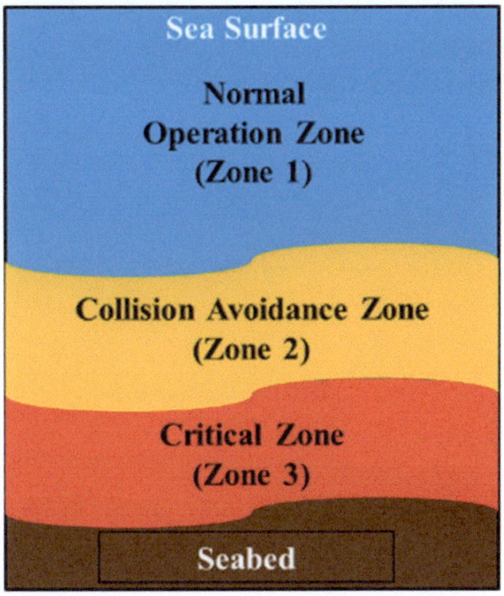

Fig. 3.17 Water column regions

Collision Avoidance Zone

This is the zone from which operation of the robot is judged to be somewhat dangerous because the robot is getting closer to the seabed. A collision avoidance maneuver based on PID control of vertical thrusters is used to smoothly stabilize the robot above a predefined critical altitude.

Critical Zone

Operation of the robot in this zone is very dangerous and presents a risk that the robot can hit the seabed, which may lead to heavy damage to the robot and especially to the ceramic transducer of the DVL. The altitude of the critical zone is adaptive and is determined based on the vertical speed of the robot and its altitude. The control program calculates the time needed to reach the seabed based on the vertical speed value. For the case in which the time to reach the seabed becomes less than a specified time limit, the thrusters are activated. This enables the robot to have enough time to decelerate the robot. The thrusters are also automatically activated in the case where the robot altitude goes below a specified critical value.

3.3.1.3 Control Priorities

The control priorities are shown in Fig. 3.18. The emergency order has the highest priority. A state of emergency is flagged when the robot battery runs below a predefined threshold. At the beginning of each mission, the mission timer is reset. When the timer reaches the maximum mission time, the emergency is activated. The emergency state can also be sent through downlink data when an abnormality is detected in the control program or in the uplink data received. Overheating, humidity, and high currents can also be added, but they are not currently implemented in the software. When the emergency is activated, the program interrupts the control algorithm and sets the neutral buoyancy to its maximum value. The

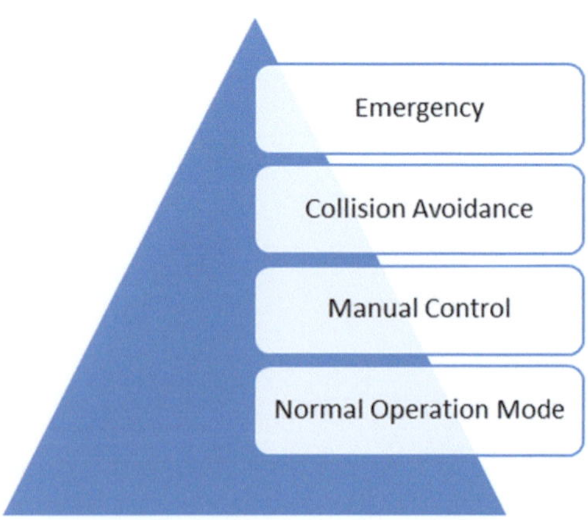

Fig. 3.18 Control priorities

data logging resumes normally and sensors are kept powered on, except in the case when the power of the system becomes too low. It is important to mention here that this emergency is a software emergency, and it is different from the emergency encountered by dropping the ballast weight. The second-place priority is the collision avoidance program. The collision avoidance uses only the thrusters. Thus, if the control program does not involve the control of the thrusters, it will keep executing. Otherwise, the thruster command from the control program will be ignored, and only the collision avoidance command will be considered. Manual control comes next. On the sea surface, only the manual control through Wi-Fi is counted, while the acoustic control is ignored. When the robot is underwater, the control through radio waves becomes impossible, and the acoustic communication is used instead. The robot is able to know its position regarding water level through the GPS status flag. When the robot is connected to Wi-Fi, the manual control can be performed either through GUI directly or through a joystick connected to the UDP port. The control of the robot through downlink data is explained in Sect. 3.2.3.3. Finally, if none of the higher-priority algorithms are flagged, the robot performs its designated operating mode described in Sect. 3.3.1.1.

3.3.2 Equations of Motion

Using the coordinate system illustrated in Fig. 3.19, where the origin of the coordinates is set at the center of gravity, the following equation of motion in the

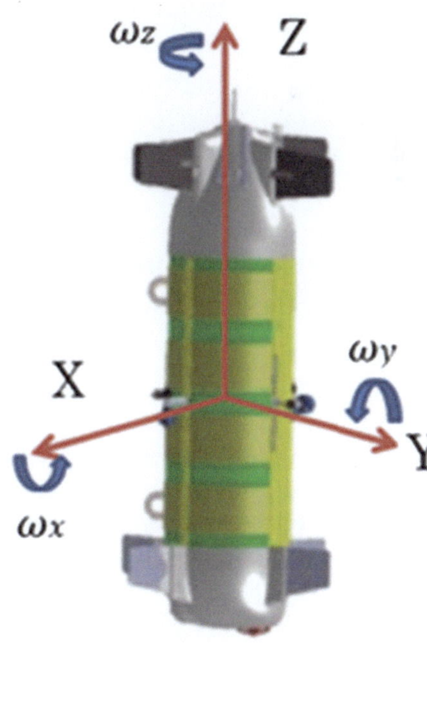

Fig. 3.19 Body-fixed coordinate system

body-fixed coordinates is expressed, where the symbols A_{11}, A_{21}, ... and I_{XX}, I_{XY}... are defined as the added mass and the moment of inertia (Azuma and Nasu 1977). Added mass is computed by the Hess-Smith method (Hess and Smith 1964) and the moment of inertia is calculated from the equipment layout. Other parameters are defined in Table 3.4.

$$\begin{pmatrix} \mathrm{M}+\mathrm{A}_{11} & 0 & 0 & 0 & \mathrm{A}_{51} & 0 \\ 0 & \mathrm{M}+\mathrm{A}_{22} & 0 & \mathrm{A}_{42} & 0 & 0 \\ 0 & 0 & \mathrm{M}+\mathrm{A}_{33} & 0 & 0 & 0 \\ 0 & \mathrm{A}_{24} & 0 & I_{XX}+A_{44} & -\mathrm{I}_{YX} & -\mathrm{I}_{ZX} \\ \mathrm{A}_{15} & 0 & 0 & -\mathrm{I}_{XY} & \mathrm{I}_{YY}+\mathrm{A}_{55} & -\mathrm{I}_{ZY} \\ 0 & 0 & 0 & -I_{XZ} & -\mathrm{I}_{YZ} & \mathrm{I}_{ZZ}+\mathrm{A}_{66} \end{pmatrix} \begin{pmatrix} \dot{U}_X \\ \dot{U}_Y \\ \dot{U}_Z \\ \dot{\omega}_X \\ \dot{\omega}_Y \\ \dot{\omega}_Z \end{pmatrix} = \begin{pmatrix} \left\{ \begin{array}{l} -F_1 \sin\Theta + F_{HX} + (M+A_{22})\,U_Y\omega_Z - A_{22}W_Y\omega_Z + A_{33}W_Z\omega_Y \\ \quad + A_{24}\omega_X\omega_Z - (M+A_{33})\,\omega_Y\omega_Z \end{array} \right\} \\ \left\{ \begin{array}{l} F_1\cos\Theta\sin\Phi + F_{HY} - (M+A_{11})\,U_X\omega_Z + (M+A_{33})\,U_Z\omega_X \\ \quad + A_{11}W_X\omega_Z - A_{33}W_Z\omega_X - A_{15}\omega_Y\omega_Z \end{array} \right\} \\ \left\{ \begin{array}{l} F_1\cos\Theta\cos\Phi + F_{HZ} + (M+A_{11})\,U_X\omega_Y - (M+A_{22})\,U_Y\omega_X \\ \quad - A_{11}W_X\omega_Y + A_{22}W_Y\omega_X - A_{24}\omega_X^2 + A_{15}\omega_Y^2 \end{array} \right\} \\ \left\{ \begin{array}{l} F_2\,(y_B\cos\Theta\cos\Phi - z_B\cos\Theta\sin\Phi) + M_{HX} + (A_{22}-A_{33})\,U_YU_Z \\ \quad + A_{33}U_YW_Z - A_{22}U_ZW_Y + A_{15}U_X\omega_Z + A_{24}U_Z\omega_X \\ \quad - A_{15}W_X\omega_Z + (I_{YY} - I_{ZZ} + A_{55} - A_{66})\,\omega_Y\omega_Z \end{array} \right\} \\ \left\{ \begin{array}{l} F_2\,(-z_B\sin\Theta - x_B\cos\Theta\cos\Phi) + M_{HY} - (A_{11}-A_{33})\,U_XU_Z \\ \quad - A_{33}U_XW_Z + A_{11}U_ZW_X - A_{24}U_Y\omega_Z - A_{15}U_Z\omega_Y \\ \quad + A_{24}W_Y\omega_Z - (I_{XX} - I_{ZZ} + A_{44} - A_{66})\,\omega_X\omega_Z \end{array} \right\} \\ \left\{ \begin{array}{l} F_2\,(x_B\cos\Theta\sin\Phi + y_B\sin\Theta) + M_{HZ} + (A_{11}-A_{22})\,U_XU_Y \\ \quad - (A_{15}+A_{24})\,(U_X\omega_X - U_Y\omega_Y) + A_{22}U_XW_Y - A_{11}U_YW_X \\ \quad + A_{15}W_X\omega_X - A_{24}W_Y\omega_Y + (I_{XX} - I_{YY} + A_{44} - A_{55})\,\omega_X\omega_Y \end{array} \right\} \end{pmatrix} \tag{3.1}$$

F_1 and F_2 are given by the following equations:

$$\boldsymbol{F_1} = \boldsymbol{\rho g V_B} - \boldsymbol{Mg} + \boldsymbol{F_B} \tag{3.2}$$

$$\boldsymbol{F_2} = \boldsymbol{\rho g V_B} + \boldsymbol{F_B} \tag{3.3}$$

The motion of the robot can be simulated by solving Eq. 3.1 using the Newmark-β method. Using this simulation, programs of guidance and control are constructed and the required times for operation are estimated.

For the rough guidance mode, a set of target points along the circles at the top and the bottom of a circular cylinder are given to perform line tracking between a point at the top of the circular cylinder and a point at the bottom of the circular

Table 3.4 Definition of parameters of the equation of motion

Symbols	Definition
(X, Y, Z)	Body-fixed coordinates
(U_X, U_Y, U_Z)	Robot velocity
(W_X, W_Y, W_Z)	Water current velocity
$(\omega_X, \omega_Y, \omega_Z)$	Angular velocities
$A_{i,j}$ $(i, j=1\sim6)$	Added mass
M	Robot mass
(X_B, Y_B, Z_B)	Buoyancy center
F_B	Buoyancy
$(F_{HX}, F_{HY,} F_{HZ})$	Hydrodynamic forces
$(M_{HX}, M_{HY,} M_{HZ})$	Hydrodynamic moments
ρ	Water density
g	Gravity acceleration
V_B	SOTAB-I's volume
$I_{i,j}$ $(i,j=X, Y, Z)$	Moment of inertia
θ	Pitch angle
ϕ	Roll angle
ψ	Azimuth angle
$\dot{()}$	Derivative with respect to time

cylinder. Let us define $(X_{E,i}\ Y_{E,i}\ Z_{E,i})$ and $(X_{E,SOTAB}\ Y_{E,SOTAB}\ Z_{E,SOTAB})$ as the target point and the position of SOTAB-I, respectively, in the Earth-fixed coordinate. If we define (X_p, Y_p, Z_p) as the offset of the present position from the target point in the body-fixed coordinate, (X_p, Y_p, Z_p) can be obtained as follows.

$$\begin{pmatrix} X_p \\ Y_p \\ Z_p \end{pmatrix} = T_B \begin{pmatrix} X_{E,i} - X_{E,SOTAB} \\ Y_{E,i} - Y_{E,SOTAB} \\ Z_{E,i} - Z_{E,SOTAB} \end{pmatrix} \tag{3.4}$$

$$T_B = \begin{pmatrix} \cos\Theta\cos\Psi & \cos\Theta\sin\Psi & -\sin\Theta \\ \cos\Psi\sin\Phi\sin\Theta - \cos\Phi\sin\Psi & \cos\Phi\sin\Theta\sin\Psi + \cos\Phi\sin\Psi & \cos\Theta\sin\Phi \\ \cos\Phi\cos\Psi\sin\Theta + \sin\Phi\sin\Psi & \cos\Phi\sin\Theta\sin\Psi - \cos\Psi\sin\Phi & \cos\Phi\cos\Theta \end{pmatrix} \tag{3.5}$$

A pair of wing angles is obtained using the PID control algorithm based on the offset components X_p and Y_p. The amount of buoyancy generated by the buoyancy control device is obtained using the PID control algorithm based on the offset Z_p.

3.3.3 Depth Control

In order to dive and perform surveying properly, SOTAB-I requires an adequate depth control. The depth control is performed by controlling the buoyancy control device. Three types of depth control methods are applied in SOTAB-I.

3.3.3.1 PID Depth Control

The first method for depth control is the PID control algorithm, which is shown in Fig. 3.20. The input of this control system is the target depth $D_{t,}$, and the output is the current depth, D. The difference between D_t and D is represented as the error e. The value of the error is input to the PID controller, which includes proportional, integral, and derivative actions. The PID controller gives a manipulated value of target buoyancy B_t as its output, which will be received by the buoyancy control device. Then the buoyancy control device will make an adjustment to the robot's buoyancy B. The results of the application of this controller in Toyama Bay experiments are discussed in Sects. 3.3.5.1 and 3.3.5.2.

Figure 3.25 in Sect. 3.3.5.1 shows that for target depth, $D_t = 300$ m, there was an approximately 130 m overshoot. However, there was a roughly 34 m overshoot for the target depth, $D_t = 150$ m, as shown in Fig. 3.28 in Sect. 3.3.5.2. The disadvantage of the PID depth control is that it is not suitable for covering long-distance dives, since this would result in a very large overshoot. This is attributed to the difficulty of tuning the parameters and the many external factors affecting performance such as temperature, salinity, and the condition of neutral buoyancy. However, the PID controller has the advantage of a fast response and good stability when it is applied around the target depth.

3.3.3.2 Depth Control with Time Estimation

The second control method is depth control with time estimation, or T_c vs. T_r control.

The CTD sensors installed in SOTAB-I enables measurement of the depth D and the vertical speed S of the robot. At every moment, it is possible to have an

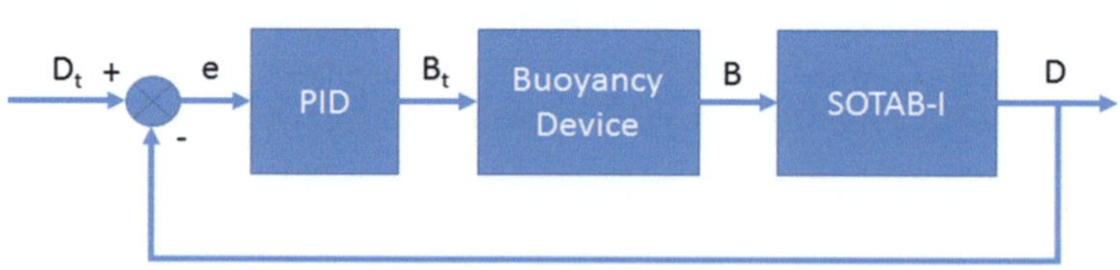

Fig. 3.20 Block diagram of PID controller for depth control

estimation of the time needed to reach the target depth T_r, at the current buoyancy level. In addition, using the buoyancy variation model based on the experimental data, and the experiments performed by SKE, the time T_c needed to change the robot buoyancy level from its current value to the neutral buoyancy condition can also be estimated.

This depth control method is based on the continuous estimation of the time to reach T_r and the time to change T_c.

- The case of estimated $T_r > T_c$: this means that it is still possible to increase the vertical speed of the robot since there will be enough time to change the buoyancy to its neutral level. Hence, the target buoyancy value is decreased.
- The case of estimated $T_r \leq T_c$: this means that there is still just enough time to change the buoyancy to the neutral level before the robot reaches its target depth. Hence, the buoyancy of the robot is increased progressively.

The time error margin T_m is also introduced in the estimation. T_m is used to compensate some error in the control mechanism of the buoyancy device. The condition for stopping increasing the robot vertical speed is when $T_r = T_c + T_m$.

This depth control method has the advantages of using the maximum buoyancy variation speed while reaching the target depth accurately. This method also allows some margin of inaccuracy in the determination of the neutral buoyancy value.

Depth control with time estimation consists of three parts:

1. Time to reach estimator

   The purpose of this part is to estimate the time needed to reach the target depth from the current depth based on the current speed of the robot.

   Input: D (m), D_t (m), and S (m/s)
   Output: T_r (s)
   Algorithm:

   - Read input parameters (D, D_t, S)
   - If ($S! = 0$) $T_r = ((D_t–D)\,/S)$
     Else $T_r = 9999$

2. Time to change estimator

   The purpose of this part is to estimate the time needed for changing the robot buoyancy level from the current level to the neutral buoyancy level.

   Input: B (%) and D (m)
   Configuration: B_n (%) and B_m (%)
   Output: T_c (s)
   Algorithm:

   - Read input parameters (B, B_n, D)
   - $R_D = D/700$
   - If ($B > B_n + B_m$) $T_c = -15.122\,(B_n–B)$

- Else if $(B < B_n - B_m)$

$$T_c = 0.0015R_D\left(B_n^3 - B^3\right) - 0.2688R_D\left(B_n^2 - B^2\right) + \left(15.725 + (34.065 - 15.725)\,R_D\right)(B_n - B)$$

- Else $T_c = 0$

R_D is the ratio used to interpolate the buoyancy model based on experimental data.

3. Buoyancy control

   The purpose of this part is to determine the target buoyancy value based on the results of the time to reach and the time to change the estimator.

   Input: B (%), T_c (s), T_r (s)
   Configuration: B_n (%), B_m (%), D_m (m) , T_m (s)
   Output: B_t (%)
   Algorithm:

   - Read input parameters (B, T_c, T_r)
   - If $(T_c < (T_r\text{–}T_m))\ B_t = B\text{–}0.3$
     Else $B_t = B + 1$
   - If $(B_t > (B_n + B_m))\ B_t = B_n + B_m$

The buoyancy error margin B_m and depth error margin D_m are introduced into the buoyancy control. They are used to compensate for some error in the control mechanism of the buoyancy device. A block diagram of the depth control with time estimation is shown in Fig. 3.21. In this figure, there is a dotted line connecting the output buoyancy B with the depth D. This indicates that the effect of buoyancy level change is to cause a change of robot depth.

The simulation of depth control with time estimation is shown in Fig. 3.22. The simulation started at a buoyancy level of 50 % and the water depth was equal to

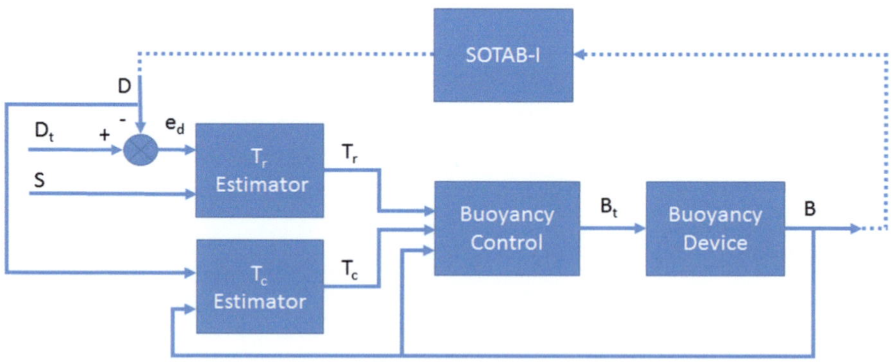

Fig. 3.21 Block diagram of PID controller for depth control

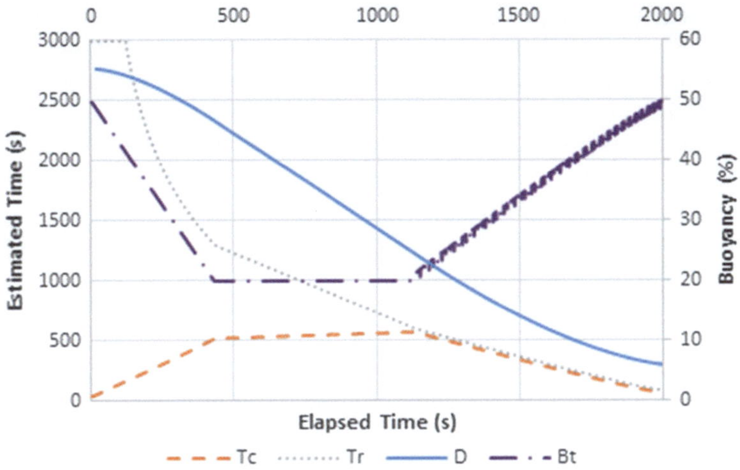

Fig. 3.22 Simulation of depth control with time estimation

0 m. With the buoyancy level equal to neutral buoyancy, then $T_c = 0$s. The value of T_r was very large because the depth was 0 m. Since $T_r > T_c$, the buoyancy control device reduced the buoyancy level. As SOTAB-I's buoyancy level was decreasing and the depth was increasing, T_c became larger and T_r became smaller. When the buoyancy level was stable at the minimum buoyancy level of 20 %, T_c increased only slightly. While at the minimum buoyancy level, SOTAB-I was diving with maximum speed, causing T_r to decrease at a faster rate than before. When $T_c \geq T_r - T_m$, the buoyancy control device started to increase the buoyancy level again. As the buoyancy level was increasing toward the neutral buoyancy level, the value of T_c was decreasing. T_r was still decreasing as the robot was still descending, but as the buoyancy level was getting closer to neutral buoyancy, the SOTAB-I descending speed became slower. Finally, the T_c and T_r were almost zero when the buoyancy level reached the neutral buoyancy level and the robot reached the target depth. Figure 3.31 in Sect. 3.3.5.3 shows the result of a sea-based experiment of SOTAB-I in Toyama Bay, where depth control with time estimation was applied. We found that SOTAB-I could reach the target depth with little overshoot.

In principle, this control method compares the time T_r needed to reach the target depth at the current buoyancy level, with the time T_c needed to change the robot buoyancy level from the current value to the neutral buoyancy level at any time during the operation. From this comparison, the buoyancy device will make an adjustment of the buoyancy level of the robot under the rule that T_c should not be larger than T_r. By doing so, SOTAB-I will always have enough time to change its buoyancy to reach the target depth. It is thus very reliable to apply the depth control with time estimation for long-distance dives. However, because the control algorithm is based on if-then-else rules, there will be oscillations when reaching the target depth. That is the reason why the PID controller is needed to achieve stability at the target depth. Hence, the combined control methods are used.

3.3.3.3 Progressive Depth Control

The third control method is progressive depth control. This method combines both of the previously explained two methods and improves their performance. It is also combined with altitude control by using DVL to detect the altitude of the robot from the seabed. This control method is very useful because sometimes SOTAB-I needs to do surveying very close to the seabed.

The progressive depth control strategy applied for the descending condition consists of four steps, as shown in Fig. 3.23, where steps 1–3 adopt the depth control with time estimation method, while step 4 adopts the PID depth control. The simulation result is shown in Fig. 3.24. In this simulation, the target depth D_t was set equal to 500 m, certain zone limit $D_{CZ} = 400$ m, depth margin $D_m = 1$ m, neutral buoyancy $B_n = 50\,\%$, and target altitude $A_t = 3$ m.

Step 1: The robot dives with a fast speed until the depth of the robot reaches the certain zone limit D_{CZ}.
The certain zone limit is the region where we are sure that the seabed will not be detected. The depth of the certain zone is input by the user on board before starting the descent. SOTAB-I will start to detect the seabed when the seabed is within the maximum detection range of the DVL, which is equal to 24 m altitude. In this first step, the depth control with time estimation scheme is used. At first, the buoyancy control device will decrease the buoyancy. After the buoyancy level becomes lower than the neutral buoyancy of the robot, SOTAB-I will start diving. The buoyancy control device will continue to reduce its buoyancy level down to 20 %, which is set as the minimum buoyancy level of the buoyancy control device, with maximum speed. Then it will increase again its buoyancy level close to the neutral buoyancy level, as shown in the step 1 in Fig. 3.24. The purpose of this strategy is that the robot should have enough time to change its buoyancy level to its neutral buoyancy when reaching the target depth. In this step, the target depth D_t is set as a fixed value, which is equal to the certain zone limit plus the DVL range R_{DVL}, minus the target altitude A_t.

Step 2: When the robot reaches the certain zone limit, the variable target depth control is started.
After passing the certain zone limit, there is a chance that the DVL will detect the seabed. Therefore, from this viewpoint, the buoyancy change is limited up to the time needed to reach the target depth T_r. In this step, the depth control with time estimation is still being used. However, the target depth D_t is set equal to the current depth D plus the DVL range R_{DVL} minus the target altitude A_t. The target depth will continuously change as the depth D of the robot decreases. Hence, it is a depth control with variable target depth, as shown in step 2 in Fig. 3.24. At this point, the buoyancy level of SOTAB-I is already close to the neutral buoyancy. Therefore, there will not be much change in the buoyancy level to ensure that the robot is able to stop when reaching the target depth, as shown in step 2 in

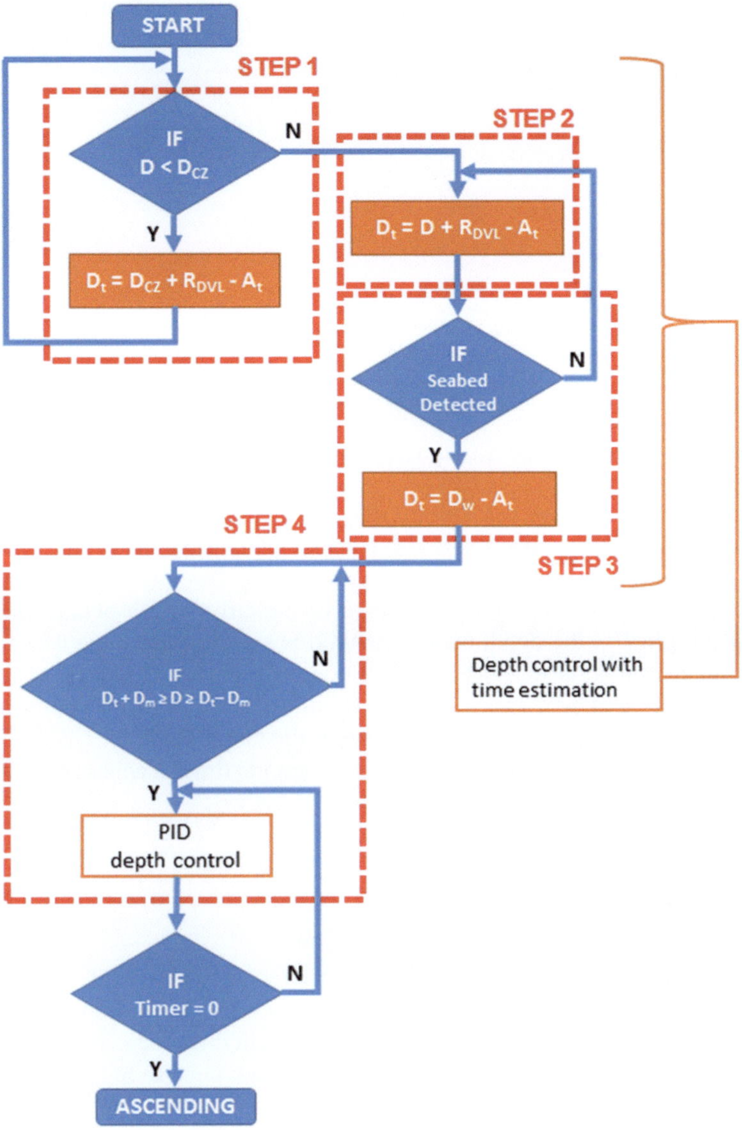

Fig. 3.23 Block diagram of progressive depth control

Fig. 3.6. As a result, the robot will dive at a steady speed, as shown in step 2 in Fig. 3.24.

Step 3: If the seabed is detected by DVL, the target depth D_t is set equal to the water depth D_w minus the target altitude A_t, as shown in step 3 in Fig. 3.23. The water depth D_w is defined as the sum of the depth D measured by CTD and the altitude

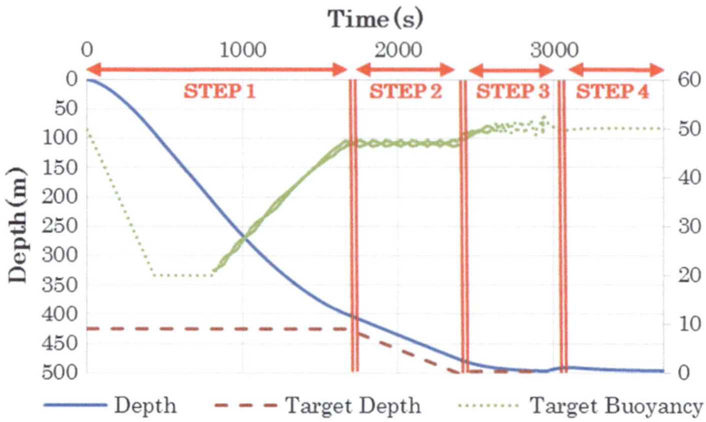

Fig. 3.24 Simulation of depth by progressive depth control

A measured by DVL. This step is also carried out by using the depth control with time estimation scheme.

Step 4: When the robot is within the range of the target depth plus or minus the depth margin D_m, the depth control method is switched from the depth control with time estimation to the PID depth control method. The depth margin D_m is usually set around 1 m as a compensation in the control mechanism of the buoyancy device. SOTAB-I will stay within the target depth for a certain period of time, which has been set on the timer. When the timer reaches zero, the robot will start ascending.

3.3.3.4 Heading Control

When performing the first stage of the survey mode, where SOTAB-I performs the water column survey by adjusting its buoyancy, it descends and gathers oceanographic data. There are two cases considered in this mode. The first one is to descend and ascend without controlling the movable wings, so SOTAB-I may drift along with the water current. The second one is to have heading control applied to control the direction of SOTAB-I during surveying. Using this heading control, it should be possible to control the trajectory of SOTAB-I during descending and ascending. It can also be used for rough guidance mode operation.

The heading control is applied for:

- X direction by multiplying the angle of wing inclination with the sine of the angle difference between the target azimuth and current yaw of SOTAB-I
- Y direction by multiplying the angle of wing inclination with the cosine of the angle difference between target azimuth and current yaw of SOTAB-I

In this case, the X direction refers to the east–west direction. The east direction is represented by positive values and the west direction is represented by negative values. The Y direction refers to the north–south direction. The north direction is represented by positive values and the south direction is represented by negative values.

The formulas for heading control are:

$$DELTX = \alpha \sin(\psi_t - \psi) \tag{3.6}$$

$$DELTY = \alpha \cos(\psi_t - \psi) \tag{3.7}$$

where α is the wing angle, ψ_t is the target azimuth, and ψ is the current azimuth (yaw), while *DELTX* and *DELTY* are the outputs for wing control.

The results of heading control will be discussed in Sect. 3.3.5.

3.3.4 Experimental Result

Field experiments of SOTAB-I have been conducted several times both in the towing tank and in the sea. In this section, only the experimental results in the towing tank in Osaka University, in Toyama Bay, and off Joetsu from 2014 to 2015 will be discussed in terms of depth control and heading control.

3.3.4.1 Field Test in Toyama Bay on the 28th of November 2014

On the 28th of November 2014, a field experiment of SOTAB-I was carried out in Toyama Bay, Japan. The experiment was performed by using a small ship of the National Institute of Technology, Toyama College, called Sazanami. The ship length is 16.0 m and its breadth is 4 m. In this experiment, the targeted depth was set as 300 m and target heading was set to 0° (north direction). The control algorithm used in this experiment was PID depth control. The performance of depth and heading control in this experiment can be seen in Figs. 3.25 and 3.26. As shown in Fig. 3.25, SOTAB-I reached 430 m depth before it started ascending. There was an approximate 130 m overshoot from the target depth. Figure 3.26 shows the orientation of SOTAB-I and Fig. 3.27 shows the horizontal position of SOTAB-I during this operation.

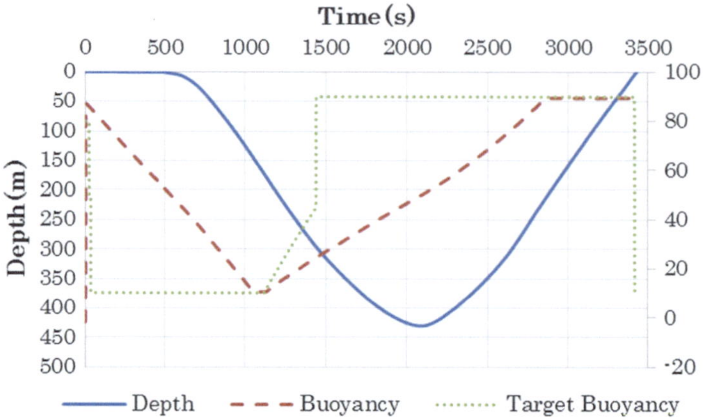

Fig. 3.25 Experimental result of depth control in Toyama Bay on the 28th of November 2014

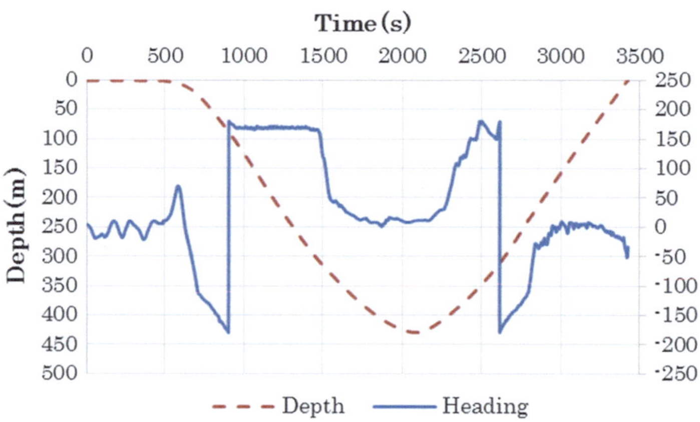

Fig. 3.26 Experimental result of heading control in Toyama Bay on the 28th of November 2014

3.3.4.2 Field Test in Toyama Bay on the 20th of March 2015

The next field experiment was also conducted in Toyama Bay, Japan, on the 20th of March 2015. The same ship, Sazanami, of the National Institute of Technology, Toyama College, was used again in this experiment. The experiment was carried out at position 36°51′N, 137°12′E with a water depth of around 700 m. In this experiment, the targeted depth was set as 150 m and target heading was set to 0° (north direction). As with the previous field test, only the PID control algorithm was used in this experiment. The performance of the depth control can be seen in Fig. 3.28. As shown in Fig. 3.28, SOTAB-I reached 184 m depth before it started ascending. There was an approximate 34 m overshoot from the target depth.

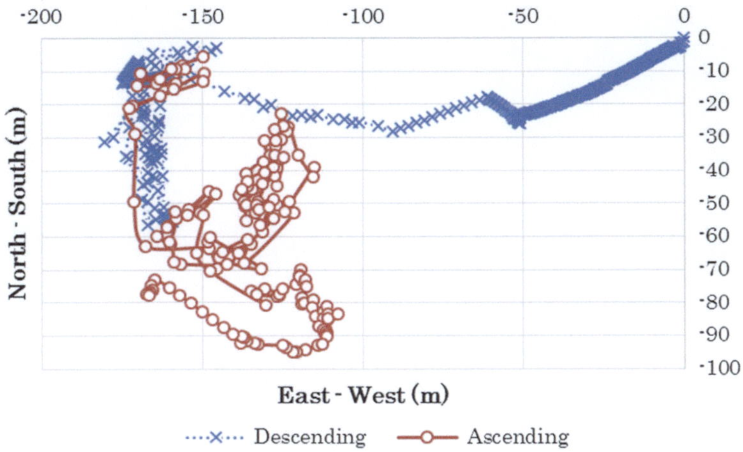

Fig. 3.27 Experimental result of horizontal positioning in Toyama Bay on the 28th of November 2014

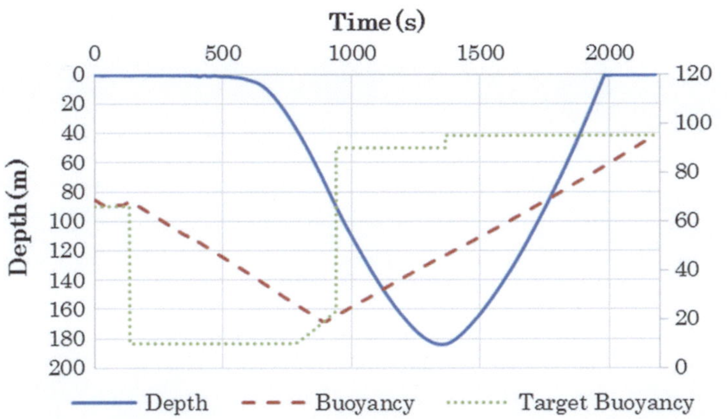

Fig. 3.28 Experimental result of depth control in Toyama Bay on the 20th of March 2015

During this experiment, the heading control was applied only when the robot was ascending, with the target azimuth 0° (north). Figure 3.29 shows the heading control performance in this experiment. When the robot started ascending, the robot's heading changed from about 50 to 0° and then was stable at around 337°. There was an approximate −23° shifting from the target azimuth. Figure 3.30 shows the horizontal position of SOTAB-I during this operation. Due to the effect of the water current, SOTAB-I moved mostly in the north and west directions.

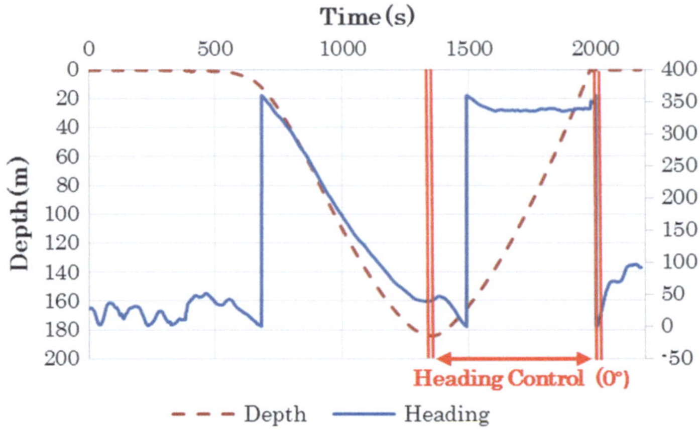

Fig. 3.29 Experimental result of heading control in Toyama Bay on the 20th of March 2015

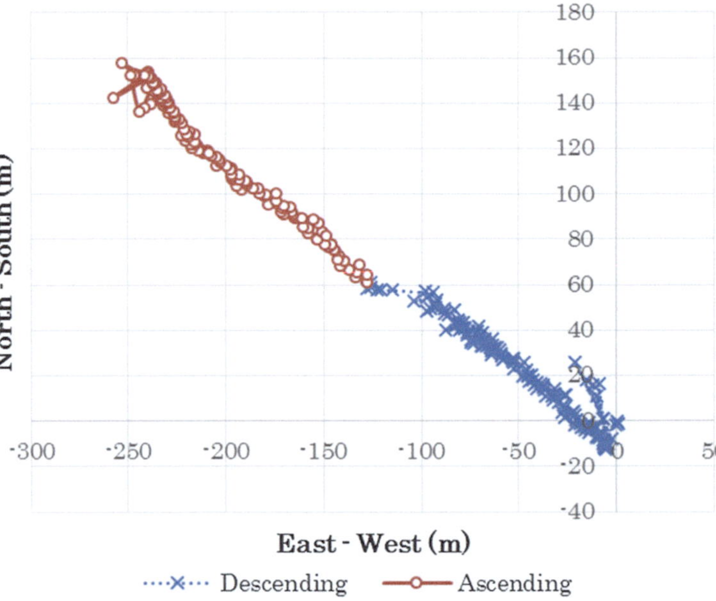

Fig. 3.30 Experimental result of horizontal positioning in Toyama Bay on the 20th of March 2015

3.3.4.3 Field Test in Toyama Bay on the 11th of June 2015

Another field experiment of SOTAB-I was carried out again in Toyama Bay, Japan, on the 11th of June 2015. However, this time the experiment was conducted using a bigger ship, the Wakashio-maru of the National Institute of Technology, Toyama

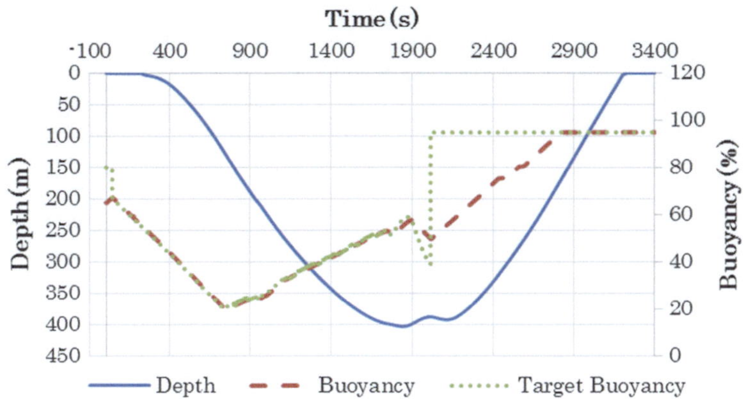

Fig. 3.31 Experimental result of depth control in Toyama Bay on the 11th of June 2015

College. The ship length is 53.59 m and its breadth is 10.0 m. The experiment site was located at 36°52′N, 137°11′E with a water depth of around 560 m.

In this sea experiment, the depth control with time estimation scheme was deployed, with a target depth of 400 m. The performance of the depth control is shown in Fig. 3.31. SOTAB-I was able to dive smoothly to the target depth of 400 m with very little overshoot. With the use of depth control with time estimation, the buoyancy device was able to adjust the buoyancy so that it could achieve neutral buoyancy when reaching the target depth. The challenge of this method is to accurately determine the value of neutral buoyancy, since there might be changes depending on the water depth and salinity. In this experiment, the timer scheme for the robot to start the ascent was not programmed in the control application; because of this, the emergency ascent mode was used to make the robot ascend. About 3 min after the robot reached the target depth, we decided to give the emergency ascent order. The order was input from the computer on board. When the emergency ascent order was given, the target buoyancy was set to 95 %, as shown in Fig. 3.31.

During this experiment, the heading control was applied only when the robot was ascending, with the target azimuth 60° (between north and west). Figure 3.32 shows the heading control performance in this experiment. When the robot started ascending, the robot's heading was already around 30°. Then it changed and stayed at approximately 35°. There was an approximate −25° shift from the target azimuth. The horizontal position of SOTAB-I during this operation is shown in Fig. 3.33. Since there was no heading control during descent, SOTAB-I drifted away along with the water current. When ascending, the robot was drifting because of the strong water current. However, the robot managed to move toward the direction between north and west eventually.

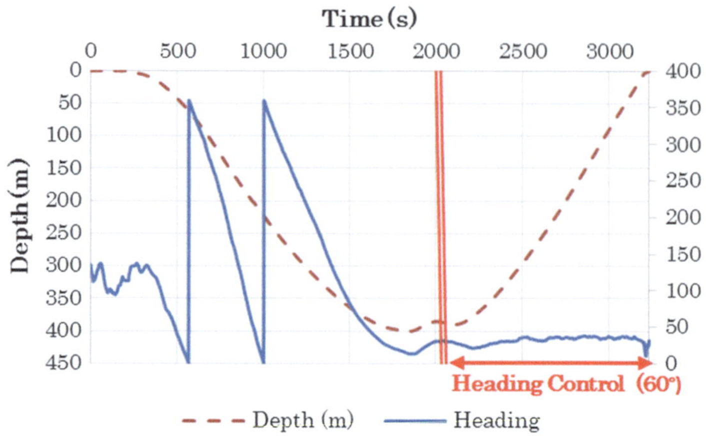

Fig. 3.32 Experimental result of heading control in Toyama Bay on the 11th of June 2015

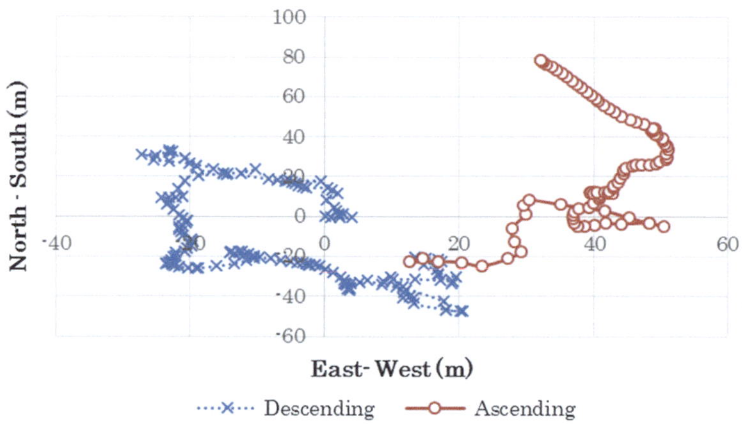

Fig. 3.33 Experimental result of horizontal positioning in Toyama Bay on the 11th of June 2015

3.3.4.4 Towing Tank Test at Osaka University on the 26th of August 2015

An experiment of the control performance of SOTAB-I was also carried out at the towing tank that belongs to the Naval Architecture and Ocean Engineering Department of Osaka University, Japan. The towing tank is 100 m in length, 7.8 m in width, and 4.35 m in depth. The purposes of this experiment were to check the whole system for the upcoming field experiments in the sea and to tune the progressive depth control.

Since the tank's depth is only 4.35 m, the second step of the progressive depth control was skipped. The target depth was set as 1.15 m, with a depth margin D_m of 0.25 m. It can be seen from Fig. 3.34 that the target buoyancy changed slightly from step 1 to step 3, which means that the control program was working. Then, the

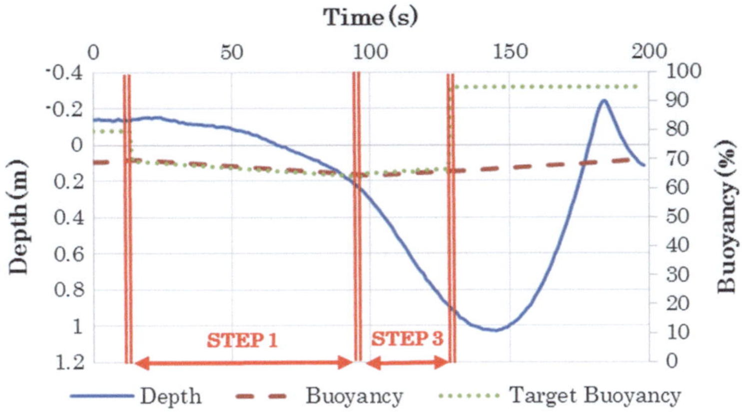

Fig. 3.34 Experimental result of depth control in Towing Tank on the 26th of August 2015

target buoyancy went to 95 % when the depth reached 0.9 m, because of the depth margin of 0.25 m from the target depth of 1.15 m.

3.3.4.5 Field Test off Joetsu, Niigata, on the 3rd of September 2015

A field test of SOTAB-I was also conducted in the sea, 20 nautical miles off Joetsu, Niigata, Japan, on the 3rd of September 2015. The experiment was carried out with use of Wakashio-maru of the National Institute of Technology, Toyama College, as the mother ship. In this experiment, SOTAB-I was equipped with the underwater mass spectrometer (UMS) to measure the methane gas concentration in the water. The targeted depth of this experiment was 700 m. Unfortunately, due to a water leakage inside the UMS, the operation was canceled midway, and SOTAB-I was commanded to come up to the surface using the emergency mode. Fortunately, some data were recovered.

The progressive depth control was deployed during this experiment. For safety reasons, the certain zone was set to 2.44 m. The target altitude was set at 3 m and the DVL range was 24 m. After reaching a water depth of 23.44 m, the control mode changed from step 1 to step 2, as shown in Fig. 3.35. As a result, SOTAB-I descended with a slow velocity. From this result, it can be seen that the progressive depth control worked well.

In this sea experiment, the heading control was applied for both descending and ascending processes. As shown in Fig. 3.36, the heading control of SOTAB-I did not work properly; it rotated all the time during descent and ascent. This might have happened because the robot was not in a balanced position due to the installation of the UMS. When installing the UMS into SOTAB-I, the dummy weight for the UMS was removed and additional buoyancy material was installed to balance the difference of weight between the UMS and the dummy weight. It is suspected that

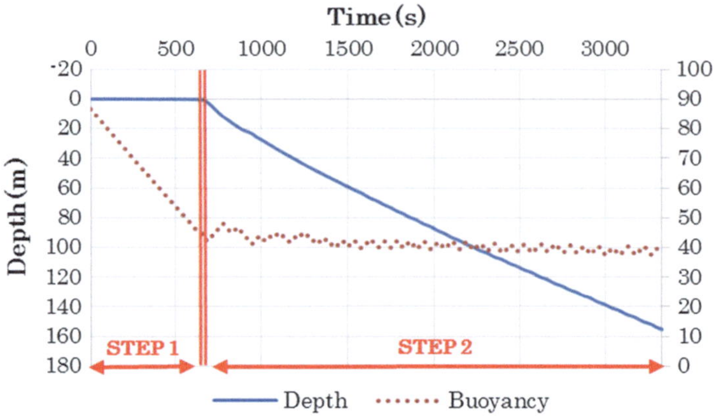

Fig. 3.35 Experimental result of depth control off Joetsu on the 3rd of September 2015

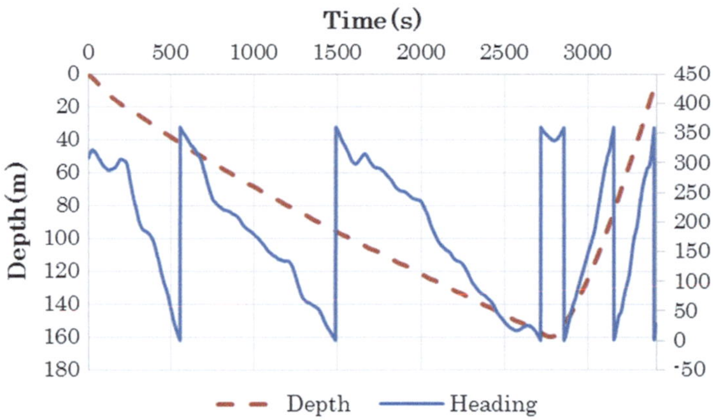

Fig. 3.36 Experimental result of heading control off Joetsu on the 3rd of September 2015

the position of the additional buoyancy material was off of the vertical centerline, causing the robot to tilt at a certain angle.

The horizontal position of SOTAB-I during this operation is shown in Fig. 3.37. Unfortunately, the USBL data obtained during this experiment was poor. Figure 3.37 shows that the horizontal position of SOTAB-I was scattered. However, this scattering of the data only occurred during descent. The data obtained during ascent were quite stable. This was also confirmed by the comparison of the depth data between the USBL and CTD, as shown in Fig. 3.38. The USBL data obtained during ascent were more stable and agreed well with the CTD depth data.

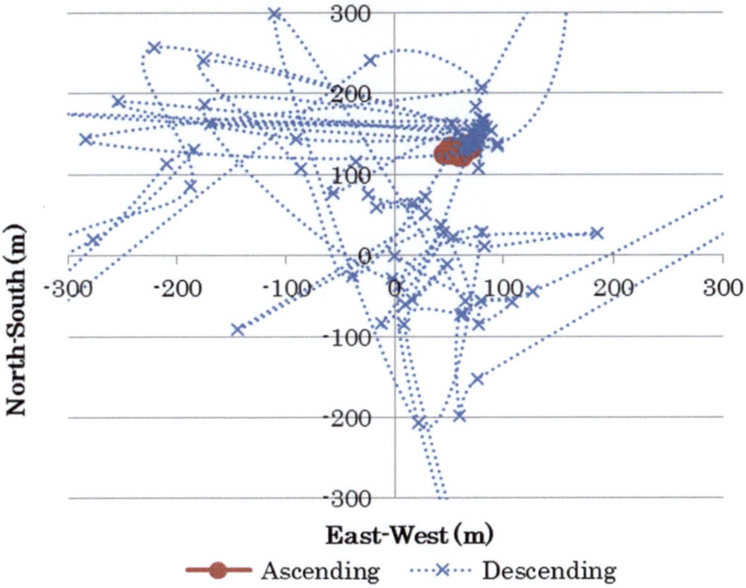

Fig. 3.37 Experimental result of horizontal positioning off Joetsu on the 3rd of September 2015

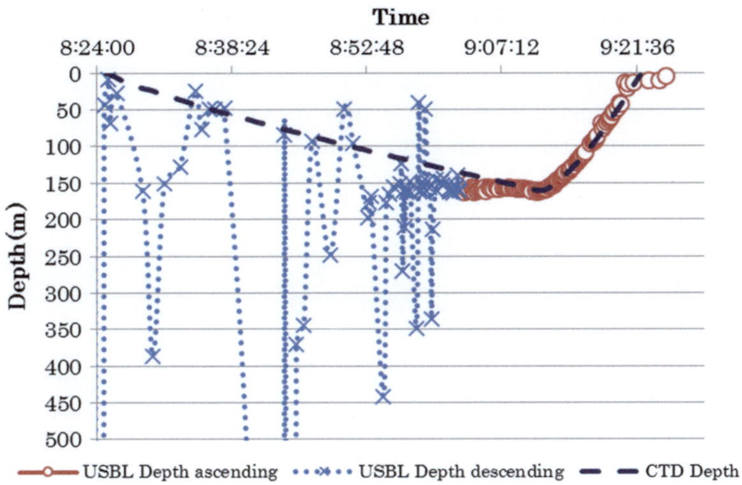

Fig. 3.38 Comparison of depth data off Joetsu on the 3rd of September 2015

3.3.5 Simulation

The simulation program was developed based on the equations of motion from Sect. 3.3.2. At first, the SOTAB-I model and the environmental model were predefined. The model of SOTAB-I considers the robot mass, added mass, length, diameter, volume, buoyancy, center of buoyancy, center of gravity, pitch, roll, and azimuthal angle, while the environmental model includes the water depth, water density, gravitational acceleration, and also at a later point, the water current velocity as input parameters.

The dynamic analysis of the equations of motion leads to the outputs of the angular velocities, the translational velocities of SOTAB-I, and the effect of the water current velocity. Furthermore, by converting from body-fixed coordinates to an Earth-fixed coordinate system, the displacement of the robot in the X, Y, and Z directions can be obtained.

In the following subsections, examples of each control simulation will be discussed.

3.3.5.1 Depth Control

The depth control simulation is performed by using the simulation program based on the equations of motion in Sect. 3.3.2. The program is used to simulate the trajectory of SOTAB-I diving in the Z axis from a starting point at the sea surface until it reaches the sea surface again. In the simulation result, the depth of SOTAB-I at every second can be obtained. The program can also simulate the buoyancy of the robot during the operation, as well as the T_c, T_r, and the velocity of SOTAB-I in the Z axis, as shown in the examples in Figs. 3.39 and 3.40, respectively. In these examples, the target depth was set to 400 m and the progressive depth control was applied.

Figure 3.41 shows the comparison of the water depth of SOTAB-I from the experimental and simulated results. In this example, the data from the field experiment in Toyama Bay on the 11th of June 2015, are used for the comparison. From this comparison, it can be concluded that the simulation of the depth control is quite accurate.

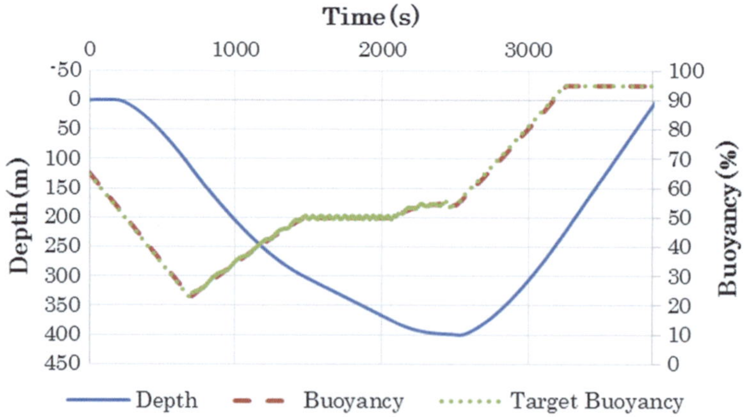

Fig. 3.39 Simulated results of buoyancy and depth in depth control

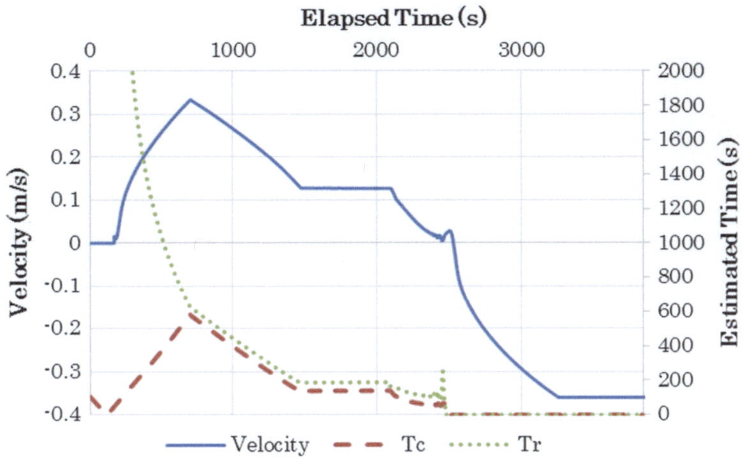

Fig. 3.40 Simulated results of *Tc*, *Tr*, and velocity in depth control

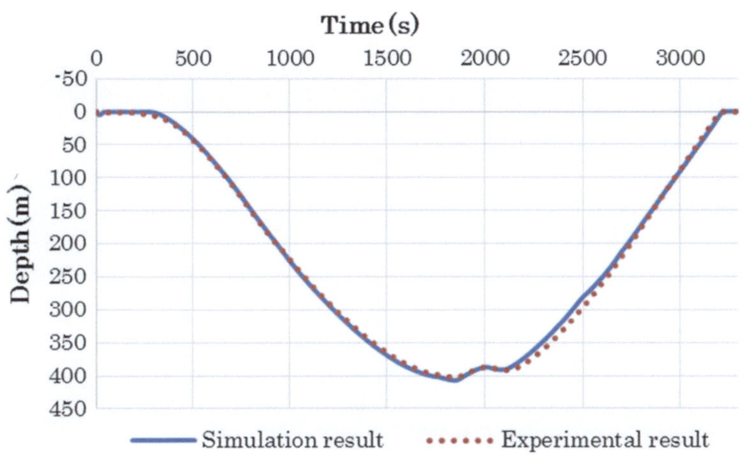

Fig. 3.41 Comparison of depth between simulation and experiment

3.3.5.2 Heading Control

The heading control simulation was performed by using the same program as the depth control simulation at the same time of the depth control simulation. The goal of this simulation was to simulate the SOTAB-I moving in the horizontal plane. Since the water current greatly affects the trajectory of the robot, the effect of the water current was considered in this simulation.

First, Fig. 3.42 shows the heading control simulation without the effects of water current taken into consideration. This figure shows the results from several simulations with different target headings. The north direction corresponds to the

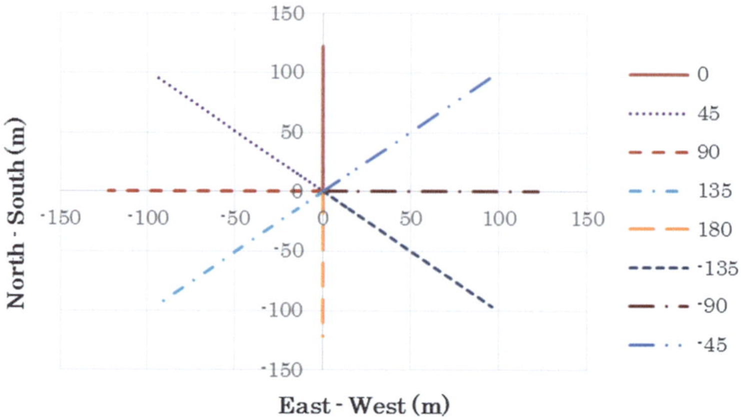

Fig. 3.42 Simulated results of heading control without water current

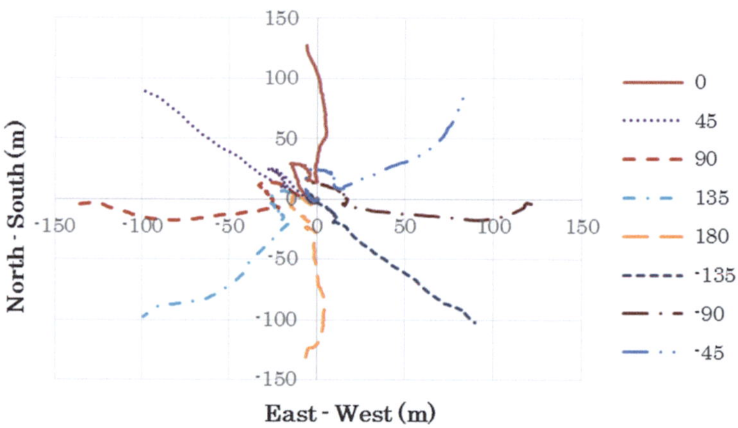

Fig. 3.43 Simulated results of heading control in water current

target heading of 0°, and then it goes counter clockwise with 90° corresponding to the westerly direction, 180° corresponding to the southerly direction, and –90° corresponding to the easterly direction.

Next, the water current was introduced into the simulation. In this simulation, a water current profile of 400 m water depth was used. The water current velocity and direction change according to depth were as shown in Figs. 3.43 and 3.44. The figures show the velocity and direction change in both east–west and north–south direction. In the shallower water depth, the water current was stronger in the north and west direction. As the depth increased, the water current became smaller and moved slightly toward the south and east directions (Fig. 3.43).

The simulated results for various target headings show that the robot was still able to move to each target heading with some curving due to the water current

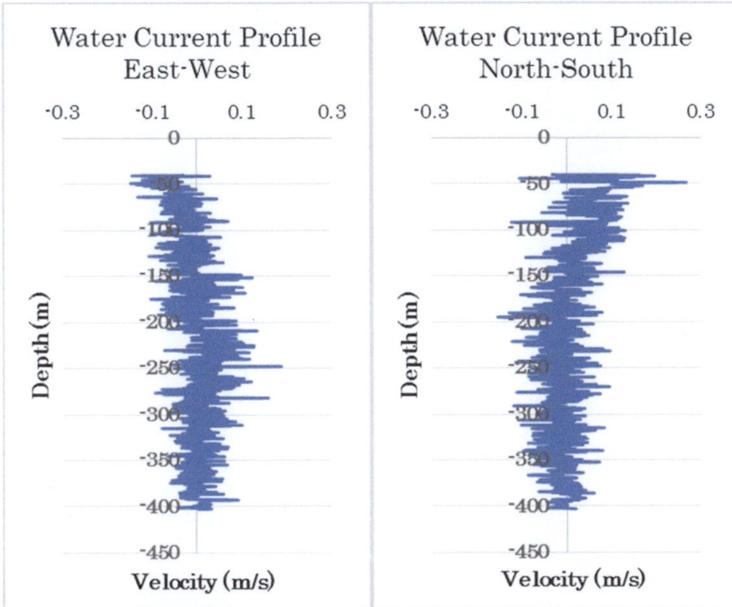

Fig. 3.44 Water current profile down to 400 m water depth

effect. Affected by the water current, the robot moved at first toward the north–west direction from the starting point (0,0). Then, as the robot descended, it changed direction toward the south–east. Since the water current velocity was smaller in the deep water, the robot was able to perform heading control, with a minimum effect from the water current, to its target heading. As the robot was ascending, it was affected by the stronger water current and moved toward the north–west.

The effect of water current on the position of SOTAB-I in the X–Z plane and in the Y–Z plane for a target heading of $-135°$ are shown in Figs. 3.45 and 3.46, respectively, for two cases of heading control: heading control during descending and ascending and heading control only during descending. The target heading of $-135°$ meant that SOTAB-I should have moved in the south–east direction. In Figs. 3.45 and 3.46, for the cases of heading control without water current, the robot moved starting from (0,0) directly toward the east (positive value) and south (negative value). We can see that the horizontal displacement during ascent was larger than that during descent. The movable wings used for the heading control are located on the top part of SOTAB-I, as shown in Fig. 3.20. These wings are directly against the water during ascent. However, during descent, the bottom part of SOTAB-I's body is directly against the water, while the movable wings are on the backside, minimizing the effect of heading control. In the case of heading control only during ascent, the robot drifted alongside the water current toward the north and west during descent in Figs. 3.45 and 3.46, respectively. In the case of

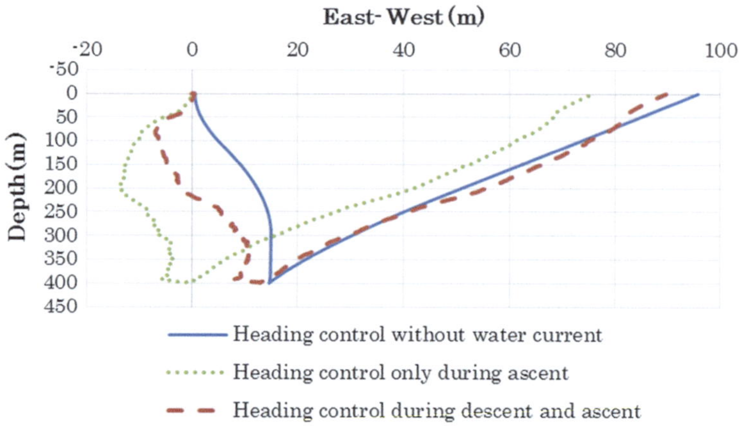

Fig. 3.45 Effect of water current on heading control in X–Z plane

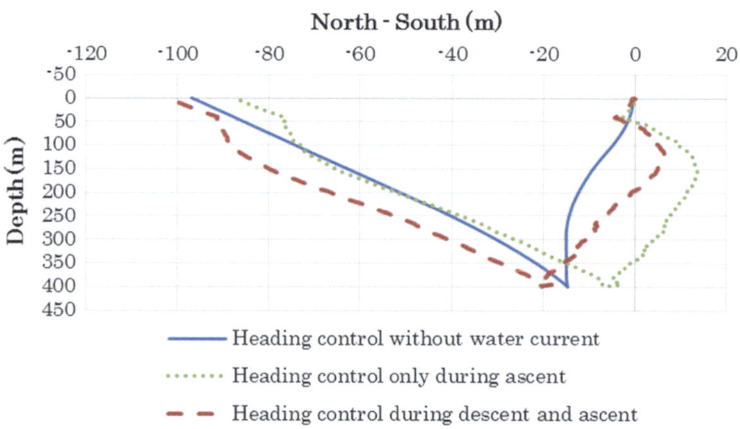

Fig. 3.46 Effect of water current on heading control in Y–Z plane

heading control during both descent and ascent, the effects of the water current on the horizontal displacement of SOTAB-I were both smaller in Figs. 3.45 and 3.46.

3.3.5.3 Simulation of Path Planning

In order to obtain the optimum results for spilled oil or methane plume surveying operation, SOTAB-I should go through these substances as much as possible. There are two main points influencing this eventuality: the trajectory of the robot coupled with spilled oil or methane plume movement in the ocean. Therefore, simulations of robot movement and spilled oil or methane plume behavior in the water (Kawahara et al. 2014) are needed. The results from both simulations will be combined to obtain

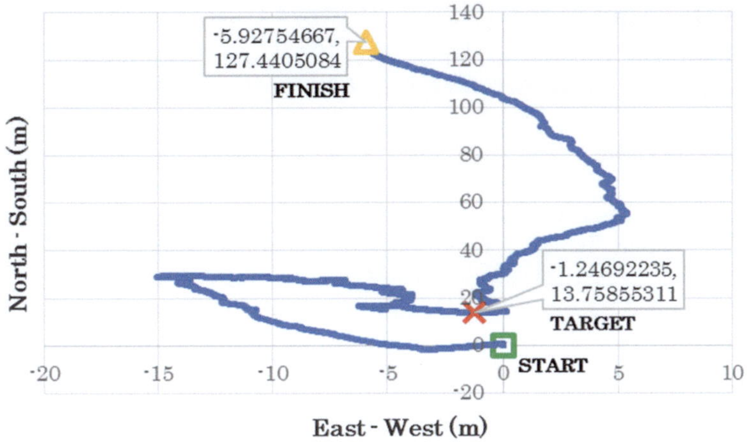

Fig. 3.47 Trajectory simulation in horizontal plane

the optimum path planning for SOTAB-I operation. Since the robot, spilled oil and methane plume movement are greatly affected by water current, it is important to obtain the vertical water current profile near the seepage point at first. This can be obtained through the initial surveying operation of SOTAB-I.

The best route for SOTAB-I in going through the spilled oil or methane plume will be determined from the simulations. After the best route for SOTAB-I is determined, the initial point for robot deployment can be obtained. This point will be used to guide the mother ship to locate the starting point for SOTAB-I deployment. From the X and Y displacements of SOTAB-I in Earth-fixed coordinates, the robot's trajectory in the horizontal plane can be acquired. While from the Z position in the Earth-fixed coordinates, the depth of SOTAB-I can be obtained.

Figure 3.47 shows the SOTAB-I trajectory simulation under the same water current conditions, as shown in Fig. 3.44. In this figure, the X and Y displacements of SOTAB-I are shown in the horizontal plane. The direction in which SOTAB-I will go under the effect of the water current can be observed from this figure. SOTAB-I starts its operation from the start point (0,0). Then, it would dive until reaching the target point. The target point is the position where the robot would reach the maximum depth. Then, it would start to ascend and resurface again at the finish point. The next step is to put the robot's trajectory onto the Earth's coordinate plane. If the target oil or methane gas seepage point is known, then the starting point where SOTAB-I should be deployed by the mother ship can also be obtained.

3.4 Water Surveying

3.4.1 Sensors Configuration and Calculation Process

3.4.1.1 CTD Data

Based on CTD measurements, it is possible to calculate the depth, salinity, density, and speed of sound. Table 3.5 summarizes oceanographic data that can be obtained with the CTD sensor with their associated symbols and scales.

Table 3.5 shows results for an example of data calculated based on CTD sensor measurements. The depth was calculated using the following formula:

$$\text{depth (m)} = \left(\left(\left(\left(-1.82\text{P}^* 10^{-15} + 2.279^* 10^{-10}\right) \text{P} - 2.2512^* 10^{-5}\right) \text{P} + 9.72659\right) \text{P}\right) / \text{g} \tag{3.8}$$

where P is the pressure in decibar and g is the local gravity value in m/s^2.

Formulas for the computation of salinity, density, and sound velocity were obtained from (Fofonoff and Millard 1983) after conversion of temperature from the international temperature scale ITS-90 to IPTS-68. Formulas for the computation of salinity, density, and sound velocity were obtained from (Fofonoff and Millard 1983).

Table 3.5 CTD-related oceanographic data

	Symbol	Unit	Comment
Temperature	T_{90}	[°C]	Given in ITS-90 scale
Conductivity	C	[S/m]	
Pressure	P	[dcb]	
Depth	D	[m]	
Salinity	S	[]	Given in practical salinity scale PSS-78
Density	ρ	[kg/m3]	Based on the equation of state for seawater – EOS80
Sound speed	V	[m/s]	Sound velocity is calculated based on Chen–Millero equation

3.4.1.2 Water Current Measurements

Introduction

Zhang and Willcox (1997) proposed a method to measure current velocities using an ADCP mounted on a cruising-type autonomous underwater vehicle (AUV) using a CTD sensor and a long baseline (LBL) acoustic navigation system. An et al. (2001) measured current velocities in shallow water using ADCP mounted on a small cruising-type AUV using a CTD sensor and an ultrashort baseline (USBL) acoustic navigation system. Stanway (2010) measured the current profile using ADCP/DVL mounted on a cruising-type AUV using a CTD sensor with bottom-lock DVL measurements. Medagoda et al. (2011) proposed an alternative approach to navigation for AUV in the mid-water column where GPS and DVL are not available by using ADCP and seafloor view-based simultaneous localization and mapping (SLAM). However, few references dealing with measurements of current profiles using ADCP and short baseline (SBL) or USBL in the mid-water column are available.

Absolute current velocity against the Earth in the mid-water column for the case of SOTAB-I equipped with ADCP/DVL, CTD, and USBL should be calculated by summing up the relative current velocity measured by ADCP on the vehicle and the speed of the vehicle, which is obtained by differentiating the position data obtained by USBL. The robot speed can also be measured directly using DVL within the device's range when it approaches the seabed. However, acoustic positioning is affected by multipath returns and circumferential noise. Thus, the accuracy of USBL data for obtaining its own speed becomes worse as depth and operation time increase. In fact, this phenomenon occurred in our experimental results. To obtain SOTAB-I's speed accurately, we propose an estimation method using ADCP and USBL. In Sect. 3.4.3, vertical water column survey part, the water column current profile at a water depth of 50 m off Komatsushima by this method was evaluated by comparing with that by the method using DVL and ADCP and that by the method using ADCP of the mother ship. The water column current profile between water depths of 15 and 50 m at depth to the bottom at 700 m at Toyama Bay by this method was also evaluated and compared with that by the method using the ADCP of the mother ship. Definitions of symbols corresponding to sensor data are shown in Table 3.6.

Table 3.6 Definitions of symbols

Symbol	Definition	Sensor
$\boldsymbol{V}_c$ ($\boldsymbol{V}_{cx}$, $\boldsymbol{V}_{cy}$, $\boldsymbol{V}_{cz}$)	Relative current velocity and components	ADCP
$\boldsymbol{V}_s$ ($\boldsymbol{V}_{sx}$, $\boldsymbol{V}_{sy}$, $\boldsymbol{V}_{sz}$)	Robot speed and components	DVL
$\boldsymbol{X}_s$, $\boldsymbol{Y}_s$, $\boldsymbol{Z}_s$	Robot position	USBL
$\boldsymbol{Z}$	Depth position of robot	CTD
$\boldsymbol{\theta}$, φ, $\boldsymbol{\psi}$	Pitch, roll, and yaw angles of the robot	ADCP, azimuth, compass

USBL Characteristics

USBL is an underwater direction and distance detection device using a broadband acoustic spread spectrum signal processing system. The transceiver on the mother ship sends a signal and the transponder attached on the top of the robot replies. The distance between the mother ship and the robot is determined by the round-trip time of sound waves and water speed, and the angle is calculated from the phase difference at the reception arrays of the USBL. Then, the relative position of the transponder against the transceiver in three-dimensional space can be calculated. USBL is also used to determine the position of the diver and installation object in the water. There are several kinds of transceiver depending on the length of acoustic communication. Our robot has adopted the transceiver called TrackLink 5000/TC5000HA. The characteristics of this transceiver are shown in Table 3.7.

The frequency zone used in acoustic positioning is between 100 Hz and 1 MHz depending on the objective. Robots, which are assumed to be used in deep water such as our robot, use a transceiver operating at a relatively low frequency. The beam width is within 120°. If the transponder gets out of place in this range, communication accuracy will become significantly worse or it will lose communication itself. An image of the position between the transceiver installed on the mother ship and the transponder installed on our robot is shown in Fig. 3.48. The transceiver on the mother ship is installed less than 3 m from the sea surface. From the characteristics of the beam width, the mother ship must be operated to keep the horizontal distance between the robot and the mother ship $\sqrt{3}$ times less than the distance of the depth of SOTAB-I. However, near the sea surface, it should keep its distance because there is a risk of collision with the mother ship if they are close to each other. For this reason, accuracy of positioning data measured by USBL is worse near the sea surface. Figure 3.49 shows a comparison of measured depths obtained by CTD and USBL. The depth data measured by USBL between 0 and 50 m has noise. When we produce water column profiles, we adopt the depth data measured by CTD as the depth of the robot because CTD has high accuracy compared with the accuracy of the USBL positioning data (see Fig. 3.49).

Table 3.7 Characteristics of the transceiver in SOTAB-I

Characteristics of the transceiver (TrackLink5000/TC5000HA)	
Positional accuracy	0.15°
Measurement distance	5000 m
Beam width	120°
Target point	8 point
Frequency used	14.2–19.8 kHz
Water depth of the pressure resistance	20 m
Dimension	26 cm × 12.6–16.0 cm (diameter)
Weight	5.2 kg
Weight in water	2.3 kg

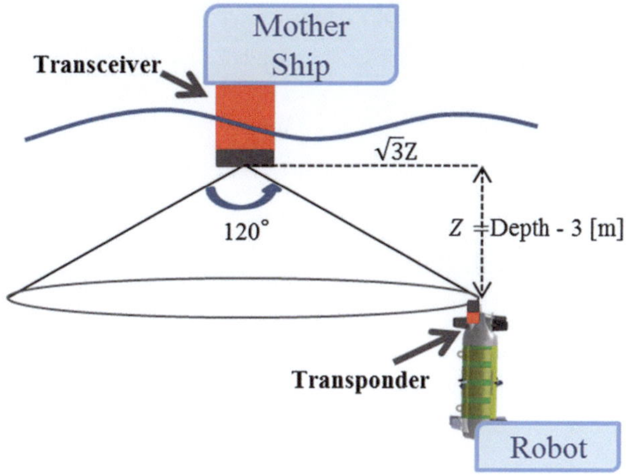

Fig. 3.48 Schematic view of the relationship between the position of the transceiver installed on the mother ship and the transponder installed on the robot

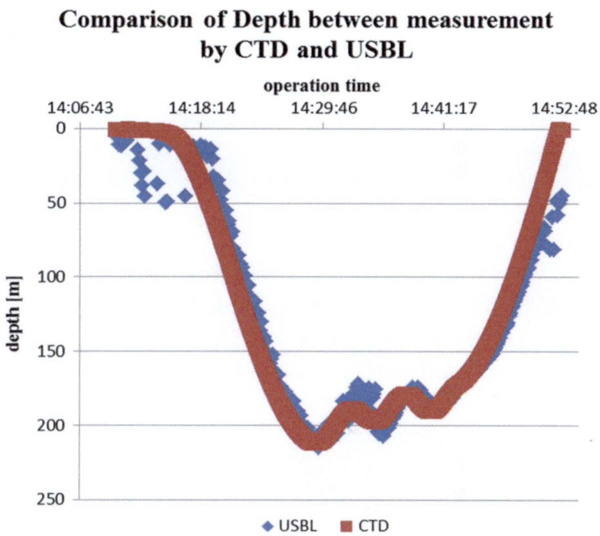

Fig. 3.49 Comparison of depths measured by CTD and USBL

ADCP Characteristics

The robot processor connects to the ADCP/DVL device through an RS232 serial port. The selected output format is PD0, which is a binary format that provides the greatest possible information. A virtual serial splitter serves to duplicate the serial data output. One is directed to serial logger software to save data in a file for later detailed analysis. The other is input to the main program for real-time processing of water currents and bottom tracking data.

The SOTAB-I configuration was set as water profiling done every second for 10 water layers and referred to also as bins with 0.5 m thickness. Measurements are configured to be given in Earth coordinates, taking into consideration tilting and bin

Table 3.8 SOTAB-I ADCP configuration

ADCP configuration	Symbol	Value
Sampling time	TE	1 s
Pings/ensemble	WP	1
Number of depth cells	WN	10
Layer thickness	WS	0.5 m
Water profiling mode	WM	1
Blank after transmit	WF	0.44 m
Salinity	ES	35
Depth of transducer	ED	0 m
1st Bin distance		0.99 m
Coordinate transformation	EX	0 × 1 F (Earth coordinates, use tilts, 3-beam solutions, bin mapping)

mapping. The most important characteristics and configuration are summarized in Table 3.8.

The ADCP is installed in the top bottom of the body. Water current data are collected when the robot is descending to reduce the turbulence that is induced by robot body motion.

Water Column Regions

The water vertical column is divided into three zones: near-surface zone, middle zone, and near-seabed zone, as shown in Fig. 3.50. The absolute water current velocity denoted as V is the sum of the absolute velocity of the robot denoted as V_s and the relative water current velocity against the robot denoted as V_c, as expressed in the following equation:

$$V = V_c + V_s \tag{3.9}$$

In the near-surface zone between 0 and 20 m of water depth, the most effective way to obtain V is to use the ADCP on the mother ship. As previously explained, the transceiver has the limit of the beam width for acoustic communication. In this zone, the horizontal distance from the mother ship to SOTAB-I is likely to become long compared with its depth. Therefore, we cannot obtain highly reliable speed data by differentiating the USBL position data. Instead, we can know those data from the ADCP on the mother ship, which can measure water currents accurately within the water depth to 100 m.

In the middle zone, the only method to estimate V_s is to differentiate the position data measured by USBL. However, the accuracy of USBL data for getting its own speed becomes worse as the water depth increases and as operational time becomes longer. In fact, these phenomena occurred in our experimental results.

In the near-seabed zone, SOTAB-I, equipped with an ADCP and a DVL, can measure both the relative current velocity and its own speed against the Earth

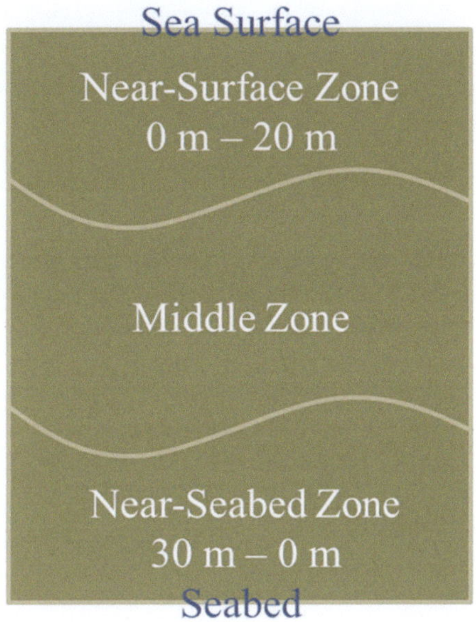

Fig. 3.50 Water layers for estimation of water current

directly. Then we can obtain absolute water currents. DVL also outputs the altitude from the seabed from 30 to 0 m.

Near-Seabed Zone

The ADCP is installed at the bottom of the robot. Water current data are collected when the robot is descending or ascending. The data during descending, however, are analyzed to reduce the influence of turbulence, which is dominantly generated in the wake region behind the fuselage of the robot during ascending. As mentioned, in this zone, the robot speed V_s can be measured by DVL, which measures the robot speed against the seabed. Thus, the relative water current velocity measured by the ADCP is the value with respect to robot-fixed coordinates (X, Y, Z). Therefore, it needs to be transformed to Earth-fixed coordinates (X_E, Y_E, Z_E). Figure 3.51 shows the relationship between robot-fixed coordinates and Earth-fixed coordinates. If we suppose V (V_X, V_Y, V_Z) as the robot speed against SOTAB-I, V_s in Earth-fixed coordinates is written as (3.10), then the absolute water current velocity can be obtained by (3.9).

$$V_s = \begin{pmatrix} V_{sx} \\ V_{sy} \\ V_{sz} \end{pmatrix} = T^{-1} \begin{pmatrix} V_X \\ V_Y \\ V_Z \end{pmatrix} \tag{3.10}$$

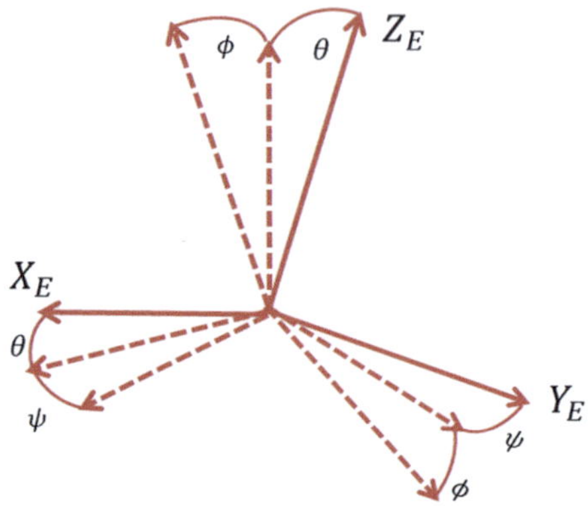

Fig. 3.51 Transformation to Earth-fixed coordinates

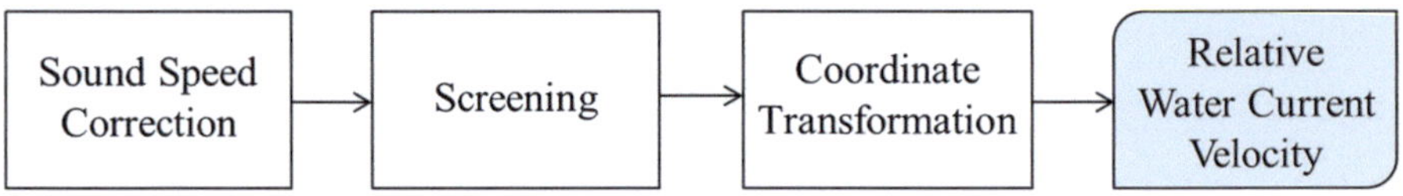

Fig. 3.52 The process of calculating relative water current velocity from ADCP

$$T = \begin{pmatrix} \cos\theta\cos\psi & \cos\theta\sin\psi & -\sin\theta \\ \cos\psi\sin\phi\sin\theta - \sin\psi\cos\phi & \sin\psi\sin\phi\sin\theta + \cos\psi\cos\phi & \sin\phi\cos\theta \\ \cos\psi\cos\phi\sin\theta + \sin\psi\sin\phi & \sin\psi\cos\phi\sin\theta - \cos\psi\sin\phi & \cos\theta\cos\phi \end{pmatrix} \tag{3.11}$$

The process of calculating relative water current velocity V_c is shown in Fig. 3.52.

The ADCP calculates the component of relative flow velocity V (m/s) for each depth cell using equation (3.12):

$$V = F_D \frac{C}{2Fs}, \tag{3.12}$$

where F_D is the measured Doppler frequency shift in the direction of each acoustic beam, C is the speed of sound in water at the transducer face in m/s, and F_S is the transmitted acoustic frequency.

From the equation, the accuracy of velocities in any coordinate system is directly connected to the sound speed: an error of 1 % in sound speed will result in 1 % error in velocity measurement. The sound speed in seawater depends on pressure, temperature, and salinity. The WHN120 integrates a thermistor to measure temperature, but it is not equipped with any pressure or salinity sensors. The ADCP calculates sound speed based on the measured temperature and preset salinity. However, the salinity of seawater is variable, especially near the sea surface. To obtain accurate velocity data, the ADCP needs to know the real speed of sound in water. For that reason, sound speed near the transducer is calculated based on the CTD sensor measurements.

It is possible to correct the velocity data in post-processing by using the following Eq. 3.13:

$$V_{\mathrm{CORRECTED}} = V_{\mathrm{UNCORRECTED}}\ (C_{\mathrm{REAL}}/C_{\mathrm{ADCP}})\,, \tag{3.13}$$

where C_{REAL} is the real sound speed at the transducer and C_{ADCP} is the sound speed used by the ADCP.

Ranges of cells, to a smaller extent, are also affected by sound speed variations and then are subject to correction. Range may be corrected by using the following equation:

$$\mathrm{L}_{\mathrm{CORRECTED}} = \mathrm{L}_{\mathrm{UNCORRECTED}}\,(C_{\mathrm{REAL}}/C_{\mathrm{ADCP}})\,, \tag{3.14}$$

where

$\mathrm{L}_{\mathrm{CORRECTED}}$: Corrected range cell location
$\mathrm{L}_{\mathrm{UNCORRECTED}}$: Uncorrected range cell location

Screening is performed automatically by the ADCP. Velocity data are subject to four kinds of screening: the correlation test, the fish-rejection algorithm, the error-velocity test, and the percent-good test. At this stage, the ADCP checks the reasonableness of the velocity components for each depth cell and flags bad data. Then, the Euler transformation must be performed using the same method as for calculating the robot speed from the DVL data (see Fig. 3.52).

Middle Zone

The ADCP data in the middle zone are treated in the same way as described above. To obtain accurate SOTAB-I speed, we suggest an estimation method using USBL positioning data.

The raw USBL data are output as a geographic coordinate system. First, we convert them to a rectangular coordinate system to differentiate position data and obtain the velocity.

Second, the distance-based outlier method (DB method) is used to remove outliers because jumping and shifting sometimes happen to the USBL position data. The DB method is an outlier method based on the distance SOTAB-I moves. It regards points as outliers if the measured values cannot be realistically achieved. The robot speed in the horizontal direction is considered to be smaller than 0.5 m/s because of its characteristic that the moving speed is not much different from the water current velocity in the horizontal direction. If the acoustic communication status between SOTAB-I and the mother ship is normal, the reception interval is every 7 or 8 s and the distance moved during that time is 4.0 m at a maximum. Then, for each measured point, we calculate the horizontal distances $R_i - 3 \sim i + 3$ between the corresponding point and each of the three previous points and the three succeeding points, as shown in Eq. 3.15. If there are more than five points within the circle with a radius of R equal to 12 m as shown in

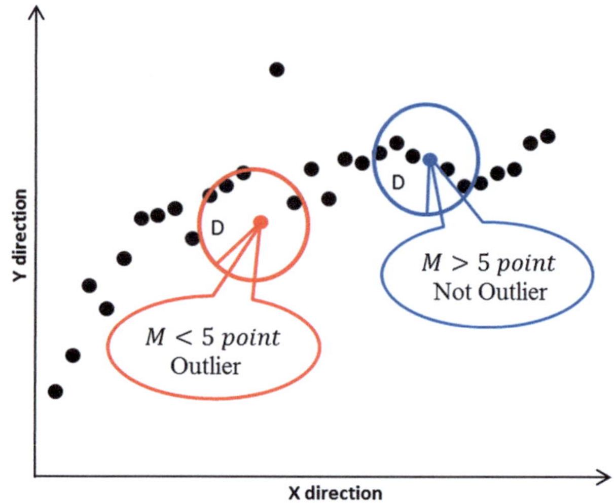

Fig. 3.53 Schematic view of the DB outlier method

Eq. 3.16, the point is determined not to be an outlier, otherwise, it is judged as an outlier. Figure 3.53 shows a schematic view of the DB outlier method.

$$R_{i-3\sim i+3} = \sqrt{(X_{i-3\sim i+3} - X_i)^2 + (Y_{i-3\sim i+3} - Y_i)^2} \tag{3.15}$$

$$R = 0.5\,[m/s] \times 8\,[s] \times 3 = 12\,[m] \tag{3.16}$$

The X, Y, and Z positioning data measured by the USBL have a correlation with the timing when the error value is generated. When one of these values is judged as an outlier, all of those are judged as outliers. An example of the effect of the DB method applied to experimental USBL data is shown in Fig. 3.54. The upper figure shows the trajectory in the east–west direction of SOTAB-I measured by USBL. The lower figure shows the trajectory in the north–south direction.

SOTAB-I usually receives USBL positioning data every 7 or 8 s by acoustic communication. However, it sometimes fails to receive data and the time interval becomes long. This phenomenon causes mistakes in the judgment in the application of the DB method, even though they are not outliers. This is the reason why we apply linear interpolation to the positioning data before treating with the DB method. Then the SOTAB-I speed can be obtained by differentiating these position data shown as Eq. 3.17:

$$\frac{X_n - X_{n-1}}{\delta t} = V_{n-1} \tag{3.17}$$

In addition, we treat one more type of outlier. Outliers are checked by whether the calculated velocity data are within $\mu \pm 3\sigma$, where μ and σ are given by:

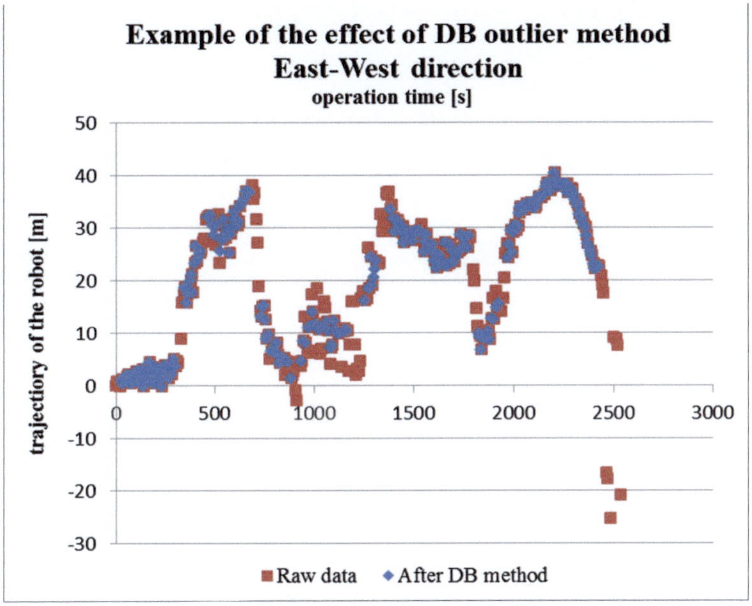

(a) An example of the effect of the DB method in east-west direction

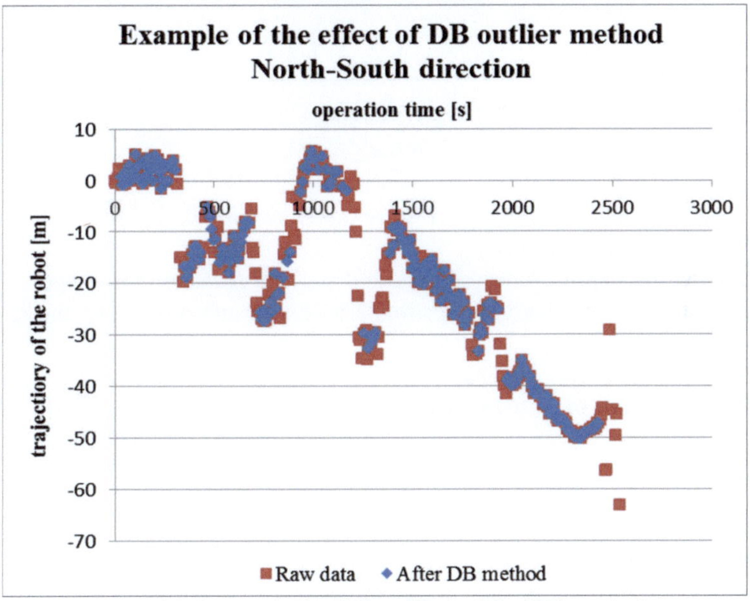

(b) An example of the effect of the DB method in north-south direction

Fig. 3.54 An example of the effect of the DB method (**a**) in east–west direction (**b**) in north–south direction

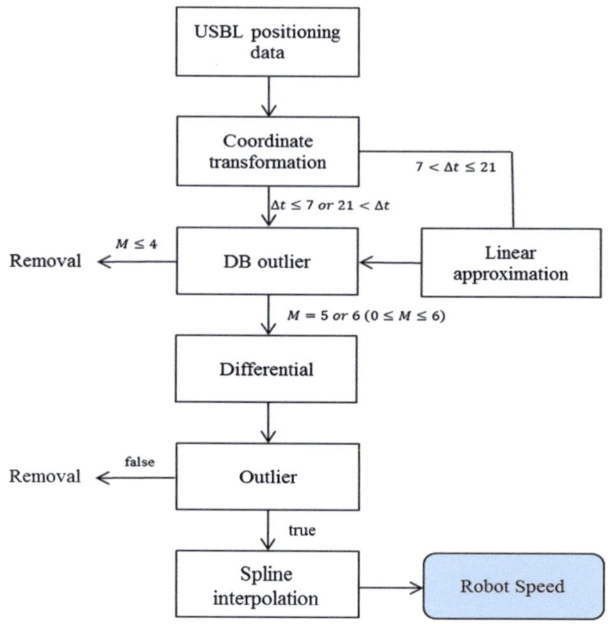

Fig. 3.55 Overview of the system diagram to estimate SOTAB-I speed from USBL

$$\mu = \frac{1}{N}\left(\sum_{i=1}^{N} V_i\right) /N, \tag{3.18}$$

$$\sigma^2 = \frac{1}{N}\sum_{i=1}^{N}(V_i - \mu)^2 . \tag{3.19}$$

N,V_i,μ, and σ denote the number of those data, data value, the average value, and standard deviation, respectively.

Finally, we use spline interpolation and get the absolute robot speed V_s. An overview of the system diagram to estimate SOTAB-I speed is shown in Fig. 3.55.

3.4.1.3 Dissolution of Substances

The UMS was calibrated for dissolved gases (methane, nitrogen, oxygen, argon, and carbon dioxide) by equilibrating acidified artificial seawater for more than 1 h with gas mixtures that contained certified mole fractions of the gases. Salinity and temperature, measured during sample analysis, allowed calculation of dissolved gas concentrations. Gas volume percentages are shown in the UMS that was calibrated for ethane, propane, and butane by equilibrating seawater with gas mixtures that contained a certified mole fraction of ethane, propane, or butane for two-point calibrations of these gases (background and one concentration). The UMS was also calibrated for VOCs by analysis of VOC standards created by serial dilution of stock solutions of benzene, toluene, and xylenes. Calibration was not performed for hydrogen sulfide or naphthalene. Each sample was analyzed until a stable signal was achieved. Blank samples (i.e., UMS residual gas backgrounds) were measured by leaving deionized water in the MIMS assembly with the sample pump inactivated overnight to allow complete degassing of the sample in contact with the membrane. The UMS assembly temperature was controlled at 25 °C during calibration to mimic deployment conditions. The UMS cast data were subsequently converted to concentrations for the dissolved gases (μmol/kg) and VOCs (ppb) from the calibration parameters and concurrently collected physical (CTD) data using algorithms and software developed by SRI International (Table 3.9).

Linear least-squares regressions provided UMS calibration coefficients for methane, nitrogen, ethane, oxygen, propane, argon, carbon dioxide, and butane concentrations using measured UMS ion currents, at m/z of 15, 28, 30, 32, 39, 40, 44, and 58. The ion current at m/z 44 (called I44), which is the mass spectrometer ion signal intensity for m/z 44 corresponding to the diagnostic ion for carbon dioxide, was also used in the nitrogen regression to account for contributions from carbon dioxide fragmentation.

Additionally, all signal intensities were background corrected by subtracting the signal intensity at m/z 5 (electronic background); this subtraction accounts for changes in electronic noise resulting from UMS temperature variability. The signal intensity at m/z 5 is used as the electronic background because there is no chemical that will give a peak in the mass spectrum at m/z 5. The "argon" or "water" correction is then used, as described in literature (Bell et al. 2007; Bell 2009), to account for temperature variations in the field. The UMS calibration parameters and deployment parameters were identical. The calibration parameters that were identical were the sample flow rate and temperature of the membrane introduction heater block. A time delay was applied to the UMS cast data to adjust for the sample travel time through the tubing and membrane permeation.

The argon and water vertical profiles are the measured ion intensities at m/z 40 (argon) and m/z 18 (water vapor) as a function of depth. These are used to

Table 3.9 Standard gas mixtures used for equilibration (in volume %)

Gas	Mixture 1	Mixture 2	Mixture 3	Mixture 4
Methane	0.0995	0.2500	2.5000	3.351
Nitrogen	Balance	Balance	Balance	Balance
Oxygen	20.85	21.0000	17.0100	9.9600
Argon	1.009	1.3010	1.0040	0.6990
Carbon dioxide	0.0990	0.7510	0.1500	0.0400

Gas	Mixture 5	Mixture 6	Mixture 7
Nitrogen	Balance	Balance	Balance
Ethane	0.1000		
Propane		0.1000	
Butane			0.1000

normalize the concentration profiles of the other analyses to account for changes in permeation through the membrane interface with increased pressure, as well as other changing environmental conditions that affect the signal intensities (Bell et al. 2007; Bell 2009; Bell et al. 2011). Therefore, high-frequency noise in these data sets was removed using a Butterworth filter prior to normalization of the other profiles.

The typical measurement accuracy at best is 2 %, but this varies for different chemicals. The response time is at best 5–10 s for the light compounds and worse for the high molecular weight compounds. A typically reasonable spatial resolution can be obtained with an ascent and descent rate of 0.5 m/s. As mentioned in the robot maneuverability section, the maximum vertical and lateral speed of the SOTAB-I are below that rate.

3.4.2 Vertical Water Column Survey in the Gulf of Mexico

At-sea experiments were performed from the 6th to the 15th of December 2013 in the Gulf of Mexico in the USA (Fig. 3.56), near where the Deepwater Horizon oil spill accident in 2010 and the Hercules 265 oil rig blowout in 2013 occurred that led to the release of methane gas. The aim of the exploration was the evaluation of the performance of the SOTAB-I's surveying abilities (Choyekh et al. 2014). Due to the strong wind and severe weather conditions, experiments were carried out in shallow water and in particular at the mouth of the Mississippi River, where the UMS data were measured. The area is characterized by its prevalent abandoned oil rigs and natural seepage of hydrocarbons (Mitchell et al. 1999).

This section is mainly focused on the experimental results obtained on the 13th of December 2013 from 13:30 to 14:30 dive.

Figure 3.57 displays the meteorological data reported by the PSTL1 station on the 13th of December 2013. The station is located in the mouth of the Mississippi River at the position (28°55′56″ N 89°24′25″ W). The distance between the PSTL1 station and the place where the experiments were carried out on the 13th of December 2013 is around 20 km. The station reported a southeastern wind direction with a gradually decreasing speed from 5 m/s at the time of the experiment. Water temperature was 13 °C, and atmospheric temperature was 20 °C.

3.4.2.1 Temperature, Salinity, and Density

Figure 3.58 displays the vertical distributions of temperature and salinity measured by the CTD. It can be observed that the sea temperature in the upper layer is colder than in the bottom layer (Fig. 3.58a), which can be explained by the cold atmospheric temperatures that dropped under 12 °C on the day that preceded the experiments. The atmospheric temperature reported by the PSTL1 buoy database was between 13 and 19 °C at the time of the experiments on the 13th of December (Fig. 3.57). The salinity level near the sea surface changed

Fig. 3.56 Gulf of Mexico experimental zone (Google Map)

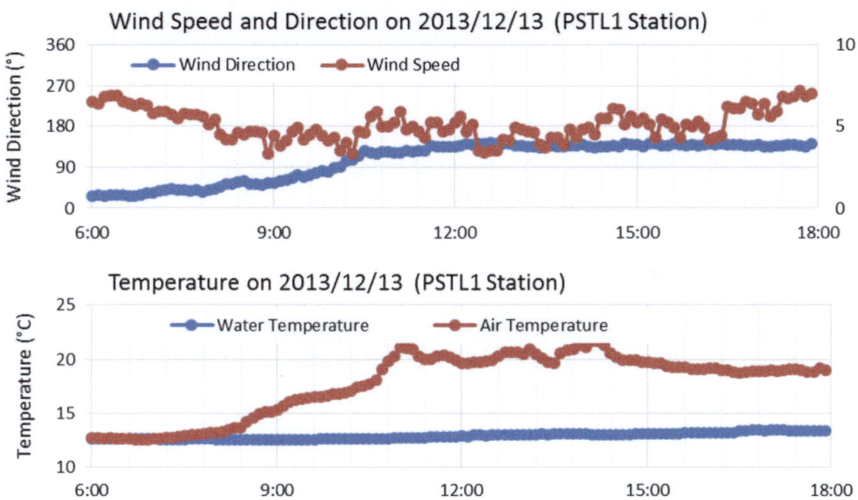

Fig. 3.57 Air and sea temperature and wind speed reported by the PSTL1 station on the 13th of December 2013

considerably from 28 ppm at 2 m water depth to 34 ppm at 8 m (Fig. 3.58b),. This can be explained by the location of the site, which is in the middle of the Mississippi mouth where freshwater flows out to the Gulf of Mexico. The freshwater layer was breached through an adjustment of the buoyancy device. The density depends on temperature, salinity, and pressure. In Fig. 3.58c, it can be observed

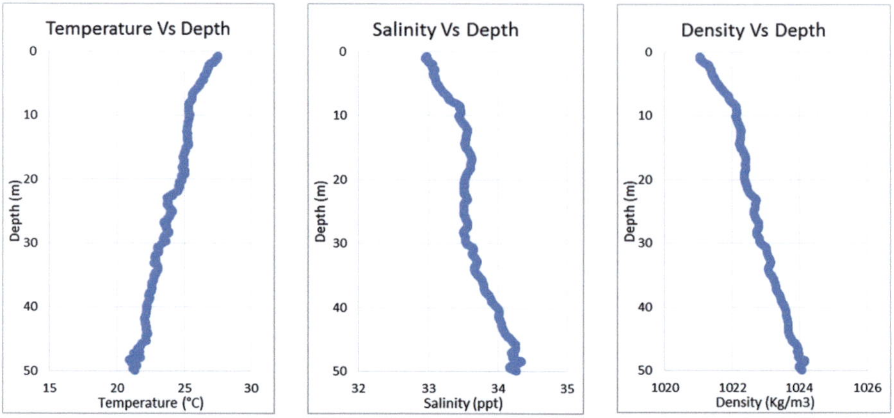

(a) Temperature (b) Salinity (c) Density

Fig. 3.58 Vertical distributions of temperature and salinity measured by the CTD: (**a**) temperature, (**b**) salinity, and (**c**) density

that the increase of density was primarily dominated by the large variation of salinity.

3.4.2.2 Water Currents

In the Gulf of Mexico experiments, the VRU was not installed, which led to less accurate positioning values due to the effect of ship tilting. For that reason, vertical water currents were only measured when the robot was within bottom track range. In this experiment, the number of layers of the ADCP was set to five layers with 0.5 m thickness. The water column profiling was based on the measurements obtained from the first layer. Figure 3.59 illustrates that the current magnitude becomes stronger in the east–west direction as the robot approaches the seabed. Water current flowed mostly in the eastern direction.

3.4.2.3 Dissolution of Substances

Figure 3.60 illustrates the change of concentration of some substances along the water column. Figure 3.60a, b shows the respective vertical concentration profiles for nitrogen and argon needed for the calculation of the other substances' dissolution profiles mentioned previously. Figure 3.60e demonstrates that the concentration of methane in the upper water layers is negligible down to a depth of 30 m, and it starts to increase steadily down to a water depth of 44.6 m. In Fig. 3.60d, it can be observed that the oxygen concentration moderately decreased from a water depth of 0 m to that of 10 m, followed by a slower rate of decline from 10 to 27 m water

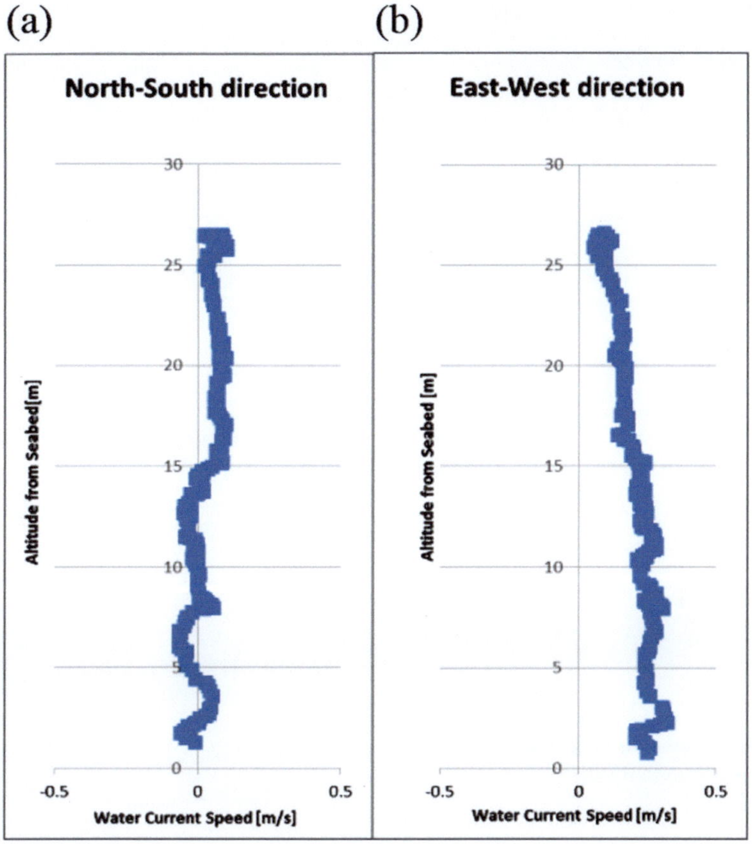

Fig. 3.59 Vertical profile of water currents (**a**) in north–south direction, and (**b**) in east–west direction

depth. Then, oxygen concentrations declined considerably from a water depth of 27 m to that of 44 m. It can be seen that the oxygen concentration decreased with increasing depth. In Fig. 3.60c, three zones can be distinguished based on the change in carbon dioxide concentrations: in water depths between 0 and 10 m, carbon dioxide concentrations decreased gradually. From 10 to 27 m, it kept decreasing, but at a slower rate. Below 30 m, carbon dioxide concentrations increased down to a water depth of 44 m.

From this perspective, we can say that the SOTAB-I succeeded in measuring dissolved substance variations along the vertical water column. Conversely, other alkanes and benzene-toluene-xylene (BTX) were below the sensory threshold and had no significant concentrations.

Very few methods will verify or corroborate the UMS measurements. We have used dissolved oxygen (DO) sensors in the past to compare the UMS oxygen measurements (m/z 32), and the comparison was generally very good (Bell et al.

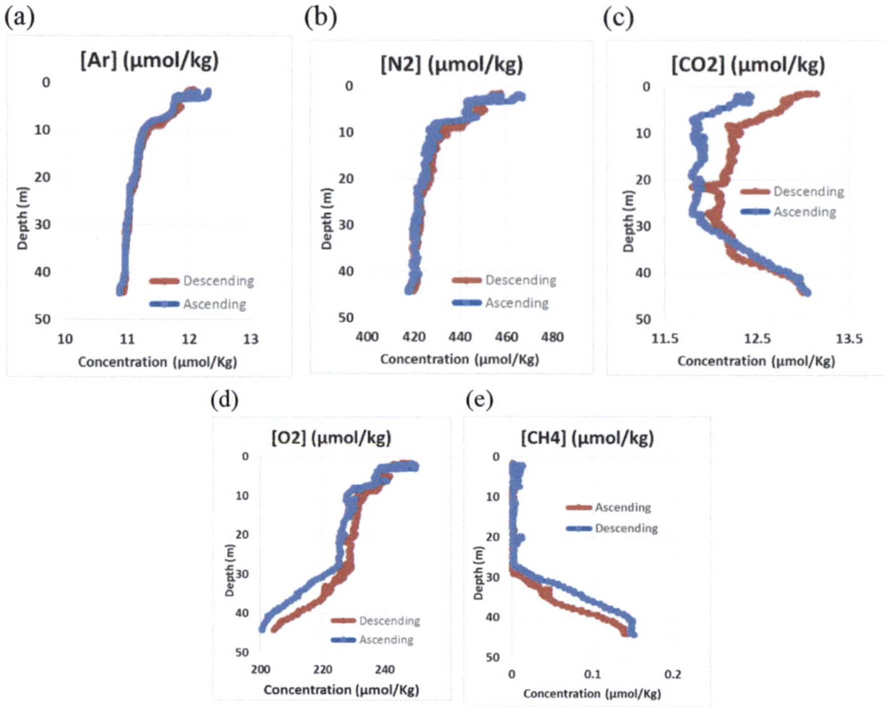

Fig. 3.60 Dissolution of substances in the water column: (**a**) methane (**b**) oxygen (**c**) carbon dioxide (**d**) nitrogen (**e**) argon

2007; Bell 2009). The SOTAB-1 deployments were not in a location where we would expect to see alkanes and BTX. We believe that the methane that we detected was biogenic methane and not associated with an oil reservoir. We have verified the UMS ability to detect these compounds in the lab and in other deployments (see Wenner et al. 2004 for BTX using an earlier version).

3.4.3 Vertical Water Column Survey in Komatsushima

Sea experiments in the shallow water region were carried out off Komatsushima, Japan (33° 96′ N, 134° 82′ E), on the 24th and 25th of July 2014, as shown in Fig. 3.61. The mother ship employed in the experiments was the "Fukae-maru" of Kobe University. The ship length is 49.97 m and its breadth is 10 m. The ship has an ADCP for measuring water currents. The water depth where the experiments took place was around 50 m.

Fig. 3.61 Pictures of the sea experiments and the location off Komatsushima

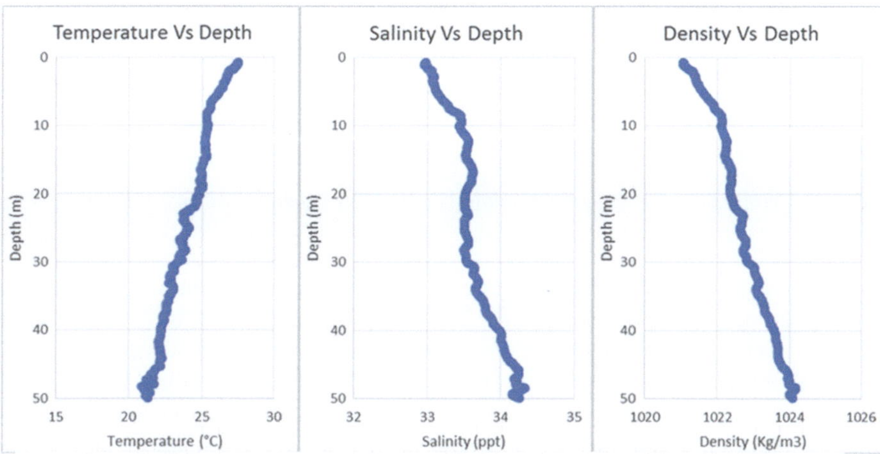

(a) Temperature (b) Salinity (c) Density

Fig. 3.62 Water column distribution of temperature, salinity, and density

3.4.3.1 Vertical Distributions of Temperature, Salinity, and Density

Figure 3.62 shows an example of data calculated based on CTD sensor measurements. As depth increases, temperature becomes lower, and salinity and density become higher.

3.4.3.2 Comparison of Vertical Water Currents Profile

We can obtain water currents at the same time and the same sea area in three ways. One is using the ADCP installed on the "Fukae-maru" and its speed. The second is using the ADCP and USBL installed on SOTAB-I. The third is using the ADCP

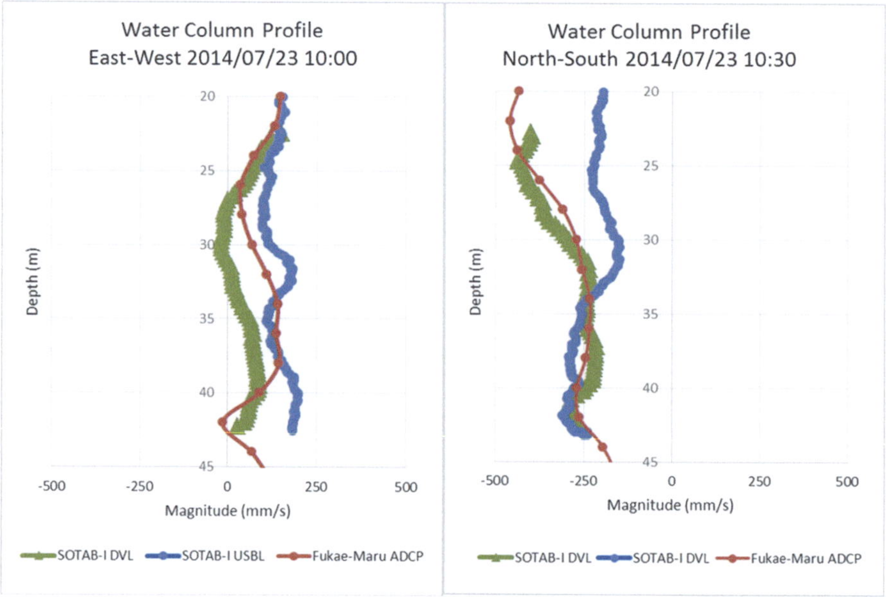

Fig. 3.63 Comparison of water current profiles measured by the three methods

and DVL installed on SOTAB-I. Figure 3.63 shows a comparison of water current profiles obtained using these methods. The profiles measured by the "Fukae-maru" are plotted along the vertical line at intervals of 3 m at a certain time. The ship drifted about 200 m from 9:50 to 10:00 while SOTAB-I was diving. Therefore, there are spatial and temporal differences of water currents between estimations and measurement by the mother ship. The results of measurements by the mother ship show almost the same curve trend regardless of a small difference in time and position. Comparison of the water current profile between estimation using DVL and measurement by the mother ship shows that the vertical profile of the water current direction is very similar, particularly in both directions. The estimation using USBL was not successful because the data measured by USBL included a lot of noise and shift in the shallow water region from 0 to 15 m as we mentioned. Comparison of the water current profile between USBL estimation and measurement by the mother ship shows that the vertical profile of the water current direction using USBL agrees well with that using the mother ship in both directions. However, the speeds are different and the maximum difference is around 0.2 m/s.

Fig. 3.64 Pictures of the sea experiments and location in Toyama Bay

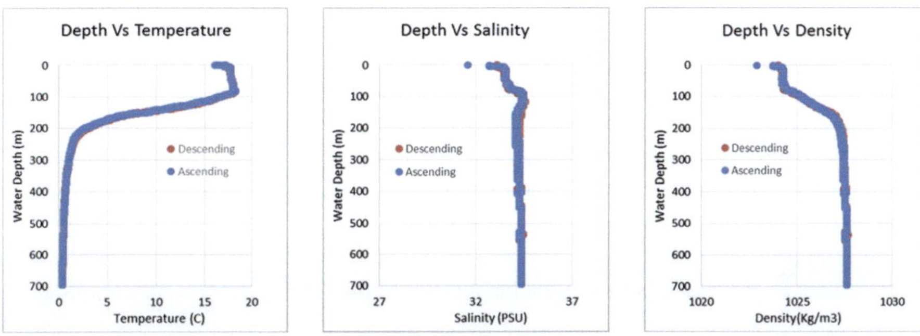

(a) Tempreture (b) salinity (c) Density

Fig. 3.65 Vertical distributions of temperature and salinity measured by the CTD: **a** temperature, **b** salinity, and **c** density

3.4.4 Vertical Water Column Survey in Toyama Bay

We conducted experiments at Toyama Bay, Japan (36° 52′ N, 137° 09′ E), twice, as shown in Fig. 3.64. The water depth where these experiments took place was around 700 m. One was carried out on the 29th and 30th of July 2014 and SOTAB-I dove to around 50 m. It indicates that DVL did not work in this region because the target depth was out of the device's range. The other was carried out on the 27th and 28th of November and SOTAB-I dove to around 700 m.

3.4.4.1 Temperature, Salinity, and Density

Figure 3.65a shows seawater temperature distribution along the water column. Three zones can be distinguished. The first, which is called "mixed layer," is very dependent on the atmospheric temperature. In the experiments day, atmospheric temperature was colder than the sea surface temperature. Seawater temperature increased with depth at a fast rate on the first 10 m, followed by a slow rate up to 90 m. In the second zone, called the thermocline zone, a sharp decrease of seawater temperature from 18 to 2° can be observed between 90 and 200 m water depth. The last zone – the deep layer – is the coldest layer and features a slow rate temperature change. The temperature reached 0.35° near 700 m water depth. The salinity profile (Fig. 3.65b) can also be divided into three distinct zones. In the surface zone, a sharp variation of salinity in the upper 10 m layer can be observed. It is due to the precipitations that preceded the experiments and to the nature of the bay, which is surrounded by the Northern Japanese Alps characterized by high mountain ranges and multiple rivers, a source of freshwater. The second zone is the halocline zone. It is characterized by a sharp increase in the salinity level up to 34.5 practical salinity units (psu) from 70 to 100 m depth. Finally, in the deep zone, a slight decrease of 34.2 psu occurred at 150 m depth, followed by a slow increase till 34.4 psu at 700 m water depth. Based on the temperature and salinity profiles, the vertical column density distribution could be obtained (Fig. 3.65c). It can be divided into three layers. In the mixed layer, low salinity water coming from rain and fresh river water is mixed with the seawater. In this layer, the density trend is more dominated by salinity variation. The second zone, called the pycnocline layer, can be distinguished between 70 and 170 m, where seawater density varies rapidly from 1,024.3 to 1,027.2 kg/m^3; it is explained by the fast increase of the salinity in the halocline zone between 70 and 100 m and then the rapid decrease of sea temperature in the thermocline zone between 90 and 200 m water depth. Finally, in the deep layer, a slow rate increase of density is observed. Density measured near 700 m water depth was equal to 1,027.6 kg/m^3. It can be noticed that the vertical water distributions of the temperature, the salinity, and the density are very similar in ascending and descending conditions. Slight differences are due to the temporal and spatial conditions in addition to the slight variance of the accuracy of the CTD sensor with its environmental temperature.

3.4.4.2 Vertical Profile of Water Currents

In the experiment on the 29th and 30th of July 2014, water currents could be obtained in two ways, estimation using USBL and ADCP measured by SOTAB-I and measurement by the mother ship "Wakashio-maru" equipped with ADCP. The length of the "Wakashio-maru" of the National Institute of Technology, Toyama College, is 53.59 m and its breadth is 10.00 m.

Figure 3.66 shows a comparison of water current profiles between these methods. The profiles measured by the "Wakashio-maru" are plotted along the vertical line at intervals of 8 m. The ship drifted around 100 m, while SOTAB-I was diving from 13:15 to 13:30. Therefore, time and position are a bit different, as in the case of the Komatsushima experiments. According to the results of the mother ship, they are almost the same curve trend regardless of time and position. The results of measurement by the mother ship show almost the same curve trend regardless of a small difference in time and position. Comparison of water current profile between estimation using USBL and ADCP of SOTAB-I and measurement by the

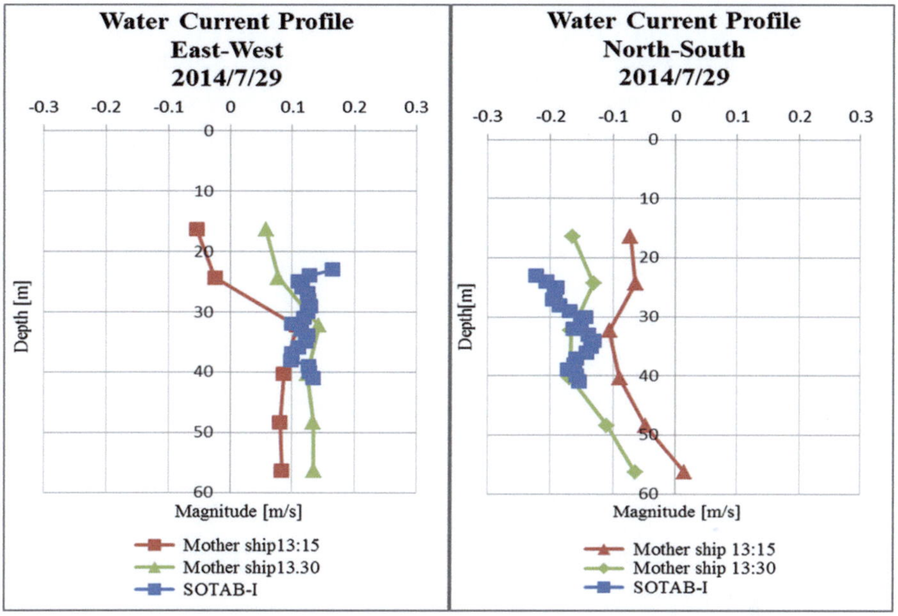

Fig. 3.66 Comparison of water current profile between estimations and measurement by the mother ship at Toyama Bay

mother ship shows that not only directions but also speeds of vertical profile of water current estimation using USBL and ADCP of SOTAB-I agree well with that using the mother ship in both directions.

The water current profile down to 700 m water depth in the experiment on the 27th and 28th of November is shown in Fig. 3.67. We used "Sazanami" as the mother ship. It is much smaller than the "Wakashio-maru" and is not equipped with ADCP. Therefore, the water current profile could not be evaluated by comparing other methods, especially the increase of the velocity from 500 m water depth in north–south direction.

3.5 Conclusions

To prevent further damage caused by oil spills and gas blowout accidents, a spilled oil and gas tracking autonomous buoy system (SOTAB-I) was developed. It has the advantages of being a compact system with extended surveying range. The robot can perform on-site measurements of several chemical substances dissolved simultaneously and is able to collect oceanographic data. The SOTAB-I can transmit data in real time with their corresponding position, making it very suitable for rapid inspection. Data collected will be processed by simulating and predicting programs

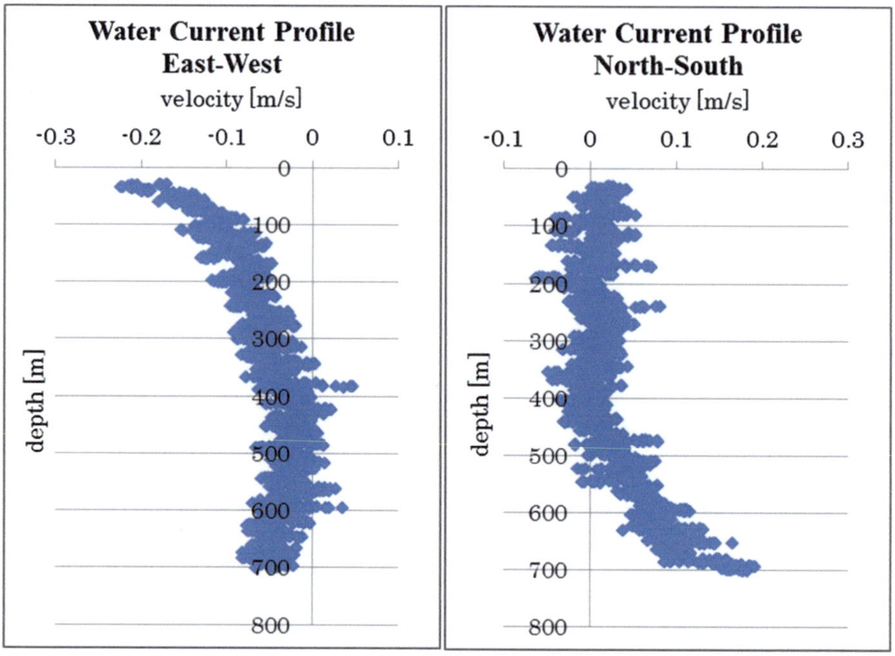

Fig. 3.67 Water current profile in deep water at Toyama Bay

that will help to explain the environmental changes due to the accident and boost the accuracy of oil drifting simulation. Consequently, establishing a better deployment strategy of collecting data becomes possible and will contribute to the efforts to avoid further damage that can be caused by oil spill disasters.

The outline of SOTAB-I and its general characteristics were described in this chapter. A summary of the main features of its internal devices was given. The SOTAB-I software design and organization were presented and an overview of its GUI was given. On the ship side, a GUI was designed, and it is a valuable tool to keep the user informed about the status of the robot and the changes around its environment. Additionally, it allows sending of commands when necessary. From the robot side, the multilayered architecture of the SOTAB-I software enabled distribution of responsibilities and enhanced the clarity of the program and its flexibility to changes. The program code optimization helped to reduce the processor usage. This contributed to the reduction in power consumption and prevented the processor from overheating inside the pressure cell. An acoustic communication data frame was established that enabled transmission of critical information to the user on the mother ship. A checksum verification method was used to secure the control of the robot through downlink, giving it an immunity against noise.

Water column regions and robot operating modes were explained. A pyramidal hierarchy was established to define each control program priority. This architecture enabled the robot to avoid dangerous situations and interrupt the control program to automatically perform software emergency ascend when a problem is detected.

The control algorithm for depth control and heading control was explained. The basic idea and characteristics of PID depth control, depth control with time estimation, and progressive depth control were described. Some simulation examples were given to give illustrate how the depth control and heading control work.

Some experimental results were shown to demonstrate the performance of SOTAB-I's control program. The Toyama Bay experiments on November 2014 and March 2015 demonstrated the performance of the depth control using only the PID control algorithm. The depth control with time estimation algorithm was applied in the Toyama Bay experiment on June 2015. The depth control performance in this experiment was much better than observed in the previous experiments.

SOTAB-I was able to reach the target depth with very small overshoot. Even though the Toyama Bay experiment on September 2015 was interrupted by the UMS leakage incident, from the collected data, we can confirm that the early stages of the progressive depth control were working.

Path planning simulation of SOTAB-I was also introduced. It is used to simulate the trajectory of SOTAB-I during operation in X, Y, and Z axes under the effect of a water current. Combined with the underwater spilled oil or methane plume behavior simulation, the best route for SOTAB-I going through the spilled oil or methane plume can be predicted. The starting point where SOTAB-I should be deployed by the mother ship can also be obtained. From this simulation, we can know the best place for SOTAB-I deployment, which allows us to find the spilled oil or methane plume faster.

A calculation method of water current velocity using SOTAB-I was proposed. The data of water current profile is essential for the prediction of spilled oil behavior.

From the Gulf of Mexico experiments in the USA, the surveying abilities of SOTAB-I in shallow water were demonstrated. The water column distribution of dissolution of substances and water currents as well as the temperature and salinity were measured. The UMS could measure the variation in the concentration of various substances simultaneously, such as methane, oxygen, and carbon dioxide. In these experiments, a vertical water column survey was conducted within 50 m water depth.

From the Komatsushima experiment and first round experiment at Toyama Bay in Japan, the oceanographic surveying abilities of SOTAB-I were extended. Not only profiles of temperature, salinity, and density, but also water column current profiles were measured, and their accuracy was shown by comparing estimated water column current profiles using DVL and ADCP and USBL and DVL with measurements by the mother ships. In the second round experiment at Toyama Bay, these efforts were continued to extend the range up to 700 m water depth.

Water column current profiles were obtained in deep water. These collected data will help to determine the environmental changes due to accidents and boost the accuracy of oil-drift simulations, which contributes to the efforts to avoid further damage that can be caused by oil spill disasters. On the other hand, Chiba et al. (2015) analyzed the characteristics of the flow pattern in Toyama Bay by onboard

ADCP observations down to 84 m water depth, finding out the anticlockwise eddy at the inner part of Toyama Bay in summer. If we combine the onboard ADCP observation and the observation of water column current profile in deep water by SOTAB-I, more detailed structure of the flow pattern can be clarified.

Appendix

Tables 3.10, 3.11, 3.12, 3.13, 3.14, 3.15, 3.16, 3.17, 3.18, 3.19 and 3.20.

Table 3.10 Characteristics of one pack lithium-ion battery

Ref	Paco KV-100 Li-Ion battery pack
Electric charge	10.0 Ah
Voltage	14.4 V
Capacity	144 Wh
Weight	1.4 Kg

Table 3.11 Processing unit specification

Ref	Advantech PCI-104, PCM-3363
Processors	Intel atom N455 single core, 1.66 GHz, Cache 512 KB
	Intel atom D525 dual core, 1.8 GHz, Cache 1 MB
Memory	1GB DDR3 800 MHz
Interfaces	USB 2.0, RS-232, RS-422/485, SMBus (configurable to I2C), Ethernet, PS/2, GPIO
Power supply voltage	5 V
Power consumption	N455: typical 7.02 W/maximum 11.825 W
	D525: typical 9.25 W/maximum 13.475 W

Table 3.12 Thruster specification

Ref	Mitsui thruster model 260
Voltage	24 V
Nominal current	12A
Thrust force	35.28 N forward, 26.46 N backward
Maximum depth	2000 m
Weight	2 Kg

Table 3.13 Main characteristics of USBL system of SOTAB-I

Ref	LinkQuest Inc. TrackLink 5000HA
Accuracy	Positioning: 0.15°
	Slant range: 0.3 m
Working range	Up to 5000 m
Operating beam width	120°
Operating frequency	14.2–19.8 kHz
Maximum depth	Transponder up to 7000 m
	Transceiver up to 20 m
Power consumption	Transmit mode: 40 W
	Receiver mode: 1 W

Table 3.14 Main characteristics of GPS receiver of SOTAB-I

Ref	Garmin GPS 15 L
Update rate	1–900 s
Acquisition times	Reacquisition: less than 2 s
	Warm: approximately 15 s
	Cold: approximately 45 s
GPS accuracy:	Position: <15 m, 95 % typical
DGPS (WAAS) accuracy	Position: <3 m, 95 % typical

Table 3.15 Main characteristics of GPS compass on the ship

Ref	Hemisphere GPS compass VS101
Update rate standard	10 Hz; optional 20 Hz (position and heading)
Horizontal accuracy	<0.6 m 95 % confidence (DGPS1)
Heading accuracy	<0.15°rms at 1.0 m antenna separation
Pitch/roll accuracy	<1°rms
Typical acquisition times	Cold start <60 s, warm start <20 s, hot start <1 s
Power consumption	~5 W nominal

Table 3.16 Main characteristics of the CTD sensor of SOTAB-I

Ref	CTD sensor SBE-49 FastCAT
Constructor	Sea-bird electronics
Sampling rate	16 samples/s
Range	Temperature: −5 to +35 °C
	Conductivity: 0–9 S/m
	Pressure: 0–7,000 m
Resolution	Temperature: 0.0001 °C
	Conductivity: 0.00005 S/m in oceanic waters
	Pressure: 0.002 % of full scale range
Weight	In air 2.7 kg, in water 1.4 kg

Table 3.17 Main characteristics of the ADCP/DVL of SOTAB-I

Ref	Navigator DVL WHN1200
Constructor	Teledyne RD instruments
Operating frequency	1200 kHz
Maximum depth	3000 m
Built-in sensors	Water temperature gauge, inclinometer, compass
Beam angle	30°
Water track velocity (ADCP)	Layer size selectable from 0.25 to 5 m
	Number of layers is selectable from 1 to 128
Bottom track velocity (DVL)	Altitude from four individual measurements
	Minimum altitude 0.5 m
	Maximum detectable altitude 30 m
	Velocity range: −/+10 m/s
	Long term accuracy: ±0.2 % ±0.1 cm/s

Table 3.18 UMS Specifications

Ref	SRI
Mass analyzer type	Linear quadrupole mass filter
Mass range	1–200 amu
Inlet system	Membrane introduction (PDMS)
Power consumption	60–80 W
Operation voltage	24 VDC
Maximum deployment time	10–14 days (exhaust limited)
Dimensions	Diameter 24 cm, length 64 cm
Weight	In air 35 kg, in water 5 kg neg.
Depth capability	2000 m

Table 3.19 Compass specifications

Ref	PNI's FieldForce TCM-XB
Accuracy	Heading 0.3°RMS
	Pitch 0.2°RMS
	Roll 0.2°RMS
Resolution	Heading <0.1°RMS
	Tilt <0.01°
Range	Heading: 0–360°
	Pitch: ±90°
	Roll: ±180°
Maximum sample rate	30 samples/s
Supply voltage	3.6–5 V
Calibration	Hard and soft iron

Table 3.20 IMU specification

Attitude and heading	
Range	Heading: roll:±180°
	Pitch: ±90°
Static accuracy	Heading: 2.0° RMS
	Pitch/roll:0.5° RMS
Angular resolution	<0.05°
Output rate	400 Hz
Gyro specifications	
Range:	±2000°/s
In-run bias stability:	<10°/h
Linearity	<0.1 % FS
Noise density	0.0035°/s $\sqrt{}$Hz
Bandwidth	256 Hz
Alignment error	±0.05°
Resolution	<0.02°/s
Accelerometer specifications	
Range	±16 g
In-run bias stability	<0.04 mg
Linearity	<0.5 °FS
Noise density	<0.14 mg/$\sqrt{}$Hz
Bandwidth	260 Hz
Alignment error	±0.05°
Resolution	<0.5 mg
Pressure sensor specifications	
Range	10–1200 mbar
Resolution	0.042 mbar
Accuracy	±1.5 mbar
Electrical	
Input voltage	4.5–5.5 V
Max power consumption	220 mW
Digital interface	Serial TTL, RS-232

References

An E, Dhanak MR, Shay LK, Smith S, Leer JV (2001) Coastal oceanography using a small AUV. J Atmos Ocean Technol 18:215–234

Azuma A, Nasu, KI (1977) The flight dynamics of an ocean space surveying vehicle. Institute of Space and Science, University of Tokyo, Tokyo, Japan. Report No. 547, Vol. 42, No. 2, pp 41–90

Bell RJ (2009) Development and deployment of an underwater mass spectrometer for quantitative measurements of dissolved gases. Ph.D. thesis, University of South Florida,St. Petersburg, Florida

Bell RJ et al (2007) Calibration of an in situ membrane inlet mass spectrometer for measurements of dissolved gases and volatile organics in seawater environ. Sci Technol 41:8123–8128

Bell RJ et al (2011) In situ determination of total dissolved inorganic carbon by underwater membrane introduction mass spectrometry. Limnol Oceanogr Methods 9:164–175. doi:10.4319/lom.2011.9.164

Chiba H et al (2015) The characteristics of the flow pattern in Toyama Bay by onboard ADCP observations -Anticlockwise eddy at the inner part of Toyama Bay in summer

Choyekh M et al (2014) Vertical water column survey in the gulf of mexico using autonomous underwater vehicle SOTAB-I. Mar Technol Soc J: Vol. 49, No. 3, 88–101

Eriksen CC et al (2001) Seaglider: a long-range autonomous underwater vehicle for oceanographic research. IEEE J Ocean Eng 26(4):424

Fofonoff N, Millard R (1983) Algorithms for computation of fundamental properties of seawater. UNESCO Tech Pap Mar Sci 44:1–53

Handa YP (1990) Effect of hydrostatic pressure and salinity on the stability of gas hydrates. J Phys Chem 94:2652–2657

Harvey J et al (2012) AUVs for ecological studies of marine plankton communities. Sea Technol 53(9):51

Hess JL, Smith AMO (1964) Calculation of non-lifting potential flow about three dimensional bodies. J Ship Res 8(2):22–44

Jakuba MV et al (2011) Toward automatic classification of chemical sensor data from autonomous underwater vehicles, intelligent robots and systems. In: IEEE/RSJ international conference on intelligent robots and systems, pp 4722–4727

Johansen Ø et al (2003) Deep spill–field study of a simulated oil and gas blowout in deep water. Spill Sci Technol Bull 8(5–6):433–443

Joye SB et al (2011) Magnitude and oxidation potential of hydrocarbon gases released from the BP oil well blowout. Nat Geosci 4:160–164

Kawahara S et al (2014) Numerical investigation on behavior of methane gas/hydrate seeping out from deep sea floor. In: The 24th ocean engineering symposium, OES24-050.

Kessler JD et al (2011) A persistent oxygen anomaly reveals the fate of spilled methane in the deep gulf of Mexico. Science 21 331(6015):312–315

Maxino TC, Koopman PJ (2009) The effectiveness of checksums for embedded control networks. IEEE Trans Dependable Secure Comput 6(1), pp 59–72

Medagoda L, Williams BS, Pizarro O, Jakuba VM (2011) Water column current profile aided localization combined with view-based SLAM for autonomous underwater vehicle navigation. In: IEEE International conference on robotics and automation, pp 3048–3055

Mitchell R et al (1999) Estimates of total hydrocarbon seepage into the Gulf of Mexico based on satellite remote sensing images. EOS Suppl 80:OS242

Roemmich D et al (2009) The argo program observing the global ocean with profiling floats. Oceanography 22(2):34–43. doi:10.5670/oceanog.2009.36

Servio P, Englezons P (2002) Measurement of dissolved methane in water in equilibrium with its hydrate. J Chem Eng Data 47:87–90

Shaffer G et al (2009) Long-term ocean oxygen depletion in response to carbon dioxide emissions from fossil fuels. Nat Geosci 2:p105–p109

Short RT et al (2006) Detection and quantification of chemical plumes using a portable underwater membrane introduction mass spectrometer. Trends Anal Chem 25(7):637–646

Solomon EA et al (2009) Considerable methane fluxes to the atmosphere from hydrocarbon seeps in the Gulf of Mexico. Nat Geosci 2:p561–p565

Stanway MJ (2010) Water profile navigation with an Acoustic Doppler current profiler. OCEANS 2010, Sydney. doi:10.1109/OCEANSSYD.2010.5603647

Vickery K (1998) Acoustic positioning systems. A practical overview of current systems. In: Proceedings of the 1998 workshop on autonomous underwater vehicles. Fort Lauderdale, FL, USA, pp 5–17

Vogel M et al (2001) Real-time deepwater current profiling system. Proc. OCEANS 2001 (MTS/IEEE) 1:269–274

Wenner PG et al (2004) Environmental chemical mapping using an underwater mass spectrometer. TrAC Trends Anal Chem 23:288–295. doi:10.1016/S0165-9936(04)00404-2

Yang DH, Xu WY (2007) Effects of salinity on methane gas hydrate system. Sci China Ser D Earth Sci 50(11):1733–1745, Springer

Zhang Y, Willcox JS (1997) Current velocity mapping using an AUV borne acoustic Doppler current profiler. In: Proceeding of the 10th International symposium on unmanned untethered submersible technology, Durham, pp 31–40

Chapter 4
Development of a Robotic Floating Buoy for Autonomously Tracking Oil Slicks Drifting on the Sea Surface (SOTAB-II): Experimental Results

S.S. Rathour, Naomi Kato, H. Senga, T. Tanabe, M. Yoshie, and T. Tanaka

Abstract After the 2010 *Deepwater Horizon* accident, environmental regulations shifted to a more goal-oriented approach that required risk management plans for controlling site-specific risks. The real-time long-term monitoring of spilled oil drifting behavior on the sea surface is essential for decreasing the risk to coastal environments posed by spilled oil. This paper describes an autonomous robotic platform or autonomous surface vehicle (ASV), propelled by wind and water currents for the long-term monitoring of spilled oil on the ocean surface. This paper also describes a sensor-based guidance, navigation, and control system for oil spill tracking by ASV in unsteady and uncertain environments. This paper makes a unique contribution to the literature in proposing a cluster-based decision-making algorithm for sailing the ASV based on a complete scanning history of the area surrounding the vehicle by the oil detection sensor. A Gaussian-based oil cluster filtering algorithm is introduced to identify the largest oil slick patch. The physical constraints of the ASV have been taken into account to allow for the computation of feasible maneuvering headings for sailing to avoid sailing upwind (i.e., in the direction from which the wind is coming). Finally, using neoprene sheets to simulate oil spills, field test experiments are described to validate the operation of the ASV with respect to oil spill tracking using a guidance, navigation, and control system based on onboard sensor data for tracking the artificial oil targets.

Keywords UV/fluorometry • Autonomous surface vehicle (ASV) • Oil spill tracking algorithm • Sail control • Rudder control

S. Rathour (✉) • N. Kato • H. Senga • T. Tanabe
Department of Naval Architecture and Ocean Engineering, Graduate School of Engineering, Osaka University, 2-1, Yamadaoka, Suita, Osaka 565-0871, Japan
e-mail: swarn@naoe.eng.osaka-u.ac.jp; kato@naoe.eng.osaka-u.ac.jp; senga@naoe.eng.osaka-u.ac.jp; Tanabe_Naoto@naoe.eng.osaka-u.ac.jp

M. Yoshie • T. Tanaka
Port & Airport Research Institute, 3-1-1, Nagase, Yokosuka 239-0826, Japan
e-mail: yoshie@pari.go.jp; tanaka_t@pari.go.jp

© Springer Japan 2017
N. Kato (ed.), *Applications to Marine Disaster Prevention*,
DOI 10.1007/978-4-431-55991-7_4

4.1 Introduction

At one point in time, an oil spill accident would make the news headlines. Most people have some means of finding out about major oil spill accidents, such as the 1991 Gulf War oil spill or the 2010 *Deepwater Horizon* oil spill. Small- and medium-sized oil spills, however, often go undiscovered and unreported. According to statistics provided by the International Tanker Owners Pollution Federation ([ITOPF] 2015), from 2000 to 2009, a total amount of 213,000 tons of oil was discharged into the sea. The impact of these spills, however, depends upon the volume and the type of oil spilled, the ambient conditions, and the sensitivity of the affected marine ecosystem and its inhabitants to the oil. These disasters can result in significant damage to the ocean environment and regional economies. Moreover, in high volume, residual spilled oil washing up along the coast can cause substantial long-term damage to the environment (Fingas and Charles 2001). Recovery from such damage is fraught with difficulty. After spilling into the ocean, oil tends to spread to form a slick of varying thickness depending on the viscosity of the spilled oil. In addition, minor differences in the winds and currents can cause the oil slick to become patchy. Oil weathering and transport is generally governed by the thickness and area of the oil slick. In the presence of wind and water currents, the velocity vector of a drifting oil spill is the product of 2–5 % of the wind velocity vector along the wind direction at a height of 10 m from the sea surface and the water current velocity vector (Fingas and Charles 2001). Over the past several decades, a concerted scientific effort has been made to estimate the drift and spread of spilled oil on the ocean surface. Timely spill detection, real-time in situ oceanographic data, and knowledge of the oil slick's location are fundamental to reducing the environmental impact of an oil spill. Oil spill sensing and drift estimation have emerged as a major component of oil spill contingency plans for the support of response decision-making, surveillance, and oil spill warning systems. Comparative studies of oil drift trajectories calculated using the Lagrangian oil drift model and trajectory of surface drifters—using iSPHERE oil spill and current tracking buoys—have produced conflicting results (Igor et al. 2012). The estimation of oil spill trend and spread using oil drift models depends on the accuracy of real-time in situ oceanographic data, such as ocean surface currents, the wind field, the position of the oil spill, and the scatter of the oil spillage. Hence, the real-time long-term monitoring of the drifting behavior of spilled oil on the sea surface is crucial for controlling the adverse impact of oil spills on coastal environments (Takagi et al. 2012; Tsutsukawa et al. 2012).

Thin oil slicks, or in cases where the oil slick is not clearly visible, require multi-sensor fusion for oil spill detection. Airborne or spaceborn sensors play an important role in oil spill response efforts (Jensen et al. 2008). However, still photography or video photography remains one of the most commonly practiced means of oil spill surveillance. Nevertheless, it must be remarked that oil shows no distinguishing spectral characteristics in the visible spectrum (Fingas and Brown 2007); therefore, making sense of the information from visual surveillance sensors depends more upon the expertise of the user than the discriminating ability of these optical sensors. Radar and aircraft are also expensive to deploy for long-duration surveillance. The X-band radar detection method can be used by a vessel for oil spill detection and tracking; however, the accuracy and efficiency of X-band radar detection in either very calm or extreme wave conditions are questionable due to the capillary wave-dampening effect (Fingas and Charles 2001). Moreover, one of the inherent weaknesses of most sensor systems is their inability to spot oil on beaches and among weeds or debris and to detect oil under certain lighting conditions (Brown and Fingas 2003). Lack of positive discrimination between oil and some backgrounds reduces the forensic value of these sensor systems. Notwithstanding, laser fluorosensors, with their own source of excitation, can be employed round the clock. Compact fluorescence lidar systems (Yamagishi et al. 2000), using charge-coupled device (CCD) cameras for imaging, are used to detect the fluorescence of substances excited by laser. This equipment can be helicopter mounted to provide images of spilled oil spreading and its classification, even in the dark. However, the remote surveillance of oceanography data via satellites and aircraft is restricted due to their temporal and geographical limitations/coverage. Helicopters, for example, cannot continuously track spilled oil because of their limited range and endurance and the need to be mindful of the safety of the crew. Drift buoys have traditionally been employed to track spilled oil (Goodman et al. 1995). Oceanographic data and slick drift information available from these robotic platforms provide valuable information for informing oil spill monitoring and response activities. Using woodchips and cottonseed to simulate oil spills, Goodman et al. (1995) carried out a series of experiment in the Gulf of Mexico during March 1994 to review the efficacy of tracking buoys for oil spill response and planning. Significant differences were observed in the trajectory of the tracking buoys and simulated oil spills. Discus-shaped buoys were found to move out of the woodchip area within a few hours after deployment, while agrosphere-like buoys were found to consistently move in a manner unlike that of the woodchips. The Orion and Novatech buoys were quite effective in following oil slicks under test conditions (Fingas 2011). The hull design of these drift buoys is inspired by oceanographic buoys (i.e., cylindrical or spherical shapes). However, the oceanographic buoys proved to be ineffective for oil spill tracking, because the interaction of the oceanographic buoy with winds is quite different than that of oil slicks (Fingas 2011; Goodman et al. 1995).

To overcome the limitations posed by drift buoys, Osaka University developed a Spilled Oil Tracking Autonomous Buoy (SOTAB), with descending and ascending procedures to guide itself to spilled oil, equipped with an oil detection sensor (i.e., senses the viscosity of water around the SOTAB) and a CCD camera (Kato et al. 2010; Yoshie et al. 2009). Although SOTAB has been shown to successfully detect and autonomously track artificial oil (i.e., neoprene gum sheets) on the ocean surface utilizing a noncontact sensor (i.e., CCD camera), there have been instances of tracking failure due to sea conditions (Yoshie et al. 2009). A second prototype, SOTAB-II, was developed with a controllable sail to exploit ambient wind conditions for tracking the spilled oil and subjected to proof-of-concept testing at sea (Senga et al. 2009, 2012). By controlling the size and direction of the

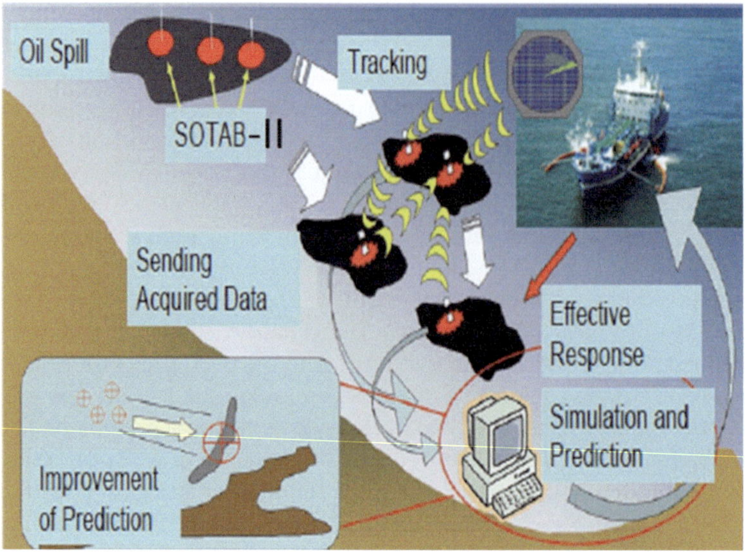

Fig. 4.1 Concept of SOTAB-II

sail as circumstances change, SOTAB-II can drift autonomously along with the spilled oil. SOTAB-II was developed with a cylindrical body to facilitate reactions to changes in the direction of the drifting oil slick. Experimental results showed that while sail control was achieved, the drift speed of SOTAB-II was unable to match the target drifting speed (i.e., resultant speed of 3 % wind speed and water current). Consequently, a yacht-shaped SOTAB-II model was proposed in order to reduce drag coefficient in water (Kato et al. 2012; Senga et al. 2013). Autonomous surface vehicles (ASVs) and autonomous underwater vehicles (AUVs) have subsequently emerged as the preferred alternative for the in situ measurement of oceanography data. An autonomous sailboat using wind and surface water current energy for propulsion and equipped with robotic platforms is a feasible long-term ocean monitoring option (Alves and Cruz 2008; Cruz and Alves 2008; Rynne and Ellenrieder 2009). The main objective of our tracking and predicting system is to allow an ASV to automatically follow a drifting oil slick and to continuously return positioning and hydrographic data to the operation base. An ASV equipped with an oil detection sensor can get closer to the oil–water interface to detect oil with greater accuracy and can continuously track the spilled oil. Such a technology, coupled with satellite and other forms of data, would facilitate the coordination of recovery operations because the data collected would better inform oil-drifting simulations, thus making it possible to predict precisely where the oil spill will travel. ASVs have the advantage of being able to track oil slicks during night when aerial surveillance of the sea surface and oil spill is impossible, thus providing a mechanism for the continuous and accurate geolocating of oil slicks (Fig. 4.1).

Since the first SOTAB-II prototype, SOTAB-II has undergone two major design changes: One is the shape hull, and the other is the shape and size of the sail, keel, and brake board; the reasons for these changes will be described in the design section below. After these design modifications, dynamic modeling and simulations were conducted to study the dynamic response of SOTAB-II (Rathour et al. 2014). These simulation results were published in proceedings of the International Society of Offshore and Polar Engineers (ISOPE) conference, 2014. The new SOTAB-II prototype comes with an oil sensor, necessitating the performance of dynamic stability testing, the results of which will be described in the design section of this paper. A decision-making algorithm for guiding SOTAB-II based on data from the oil sensor was constructed, and simulation studies were conducted to study the behavior of SOTAB-II. For this simulation, and due to its simplicity, the Lehr et al. (1984) empirical oil spill drift and spread model was adopted. Oil slick diffusion and weathering was not accounted for during the simulation stage. The decision-making algorithm generated target headings and target speeds every 20 s. After hardware and software installation, it was found that the oil sensor takes two seconds to rotate to a new target point and one second to take the target point reading. Consequently, the update time of the decision-making algorithm was modified to three seconds. The oil sensor-based decision-making algorithm also considers the time history of the scanned area around SOTAB-II to deduce the target's heading and speed (Rathour et al. 2015a, b). The decision-making algorithm considered the minimum value within the time history of the scanned area for target heading derivation and generates a target heading in the direction diametrically opposite the target point with the minimum time history value. To find the optimal value of the target heading and to deal with the possibility of weathering which can lead to the breakup of the oil slick into multiple clusters around SOTAB-II, the decision algorithm was upgraded. The updated algorithm includes the time history of the oil sensor dataset and the current sensor data to derive the target heading and speed. Based on current sensor data, the decision-making algorithm determines the number of oil slick clusters surrounding SOTAB-II and filters out the clusters based on user requirements.

In this study, a Gaussian-based filter has been used to filter out the required cluster. Neoprene sheets are used for the simulated oil spill. Based on the experimental outcomes, it was decided to track the largest patch of neoprene sheets as the sheets were found to disperse easily. Also, as we are using neoprene sponge rubber for the simulated oil spill, tracking and following a large patch is easy because the oil sensor can easily predict the continuity of the simulated oil spill and the individual neoprene sheets after they interact with the hull of SOTAB-II and begin to drift away. Therefore, the objective of the filter was to filter out the cluster with the highest value of detected time history. The cluster filtering algorithm can easily be modified to filter out newborn slicks due to weathering and breaking up of oil slicks. This could be exploited to encircle the oil spill using multiple SOTAB-II, thus providing real-time oil slick spread data.

The main contributions of the paper are as follows:

1. A generalized cluster-based decision-making algorithm capable of easily deducing a target's heading, even in situations where SOTAB-II finds itself surrounded by multiple small patches of oil slick
2. A discrete Gaussian function to filter out the desired oil spill cluster surrounding SOTAB-II
3. Sail and rudder trimming using a proportional–integral–derivative (PID) controller

4.2 SOTAB-II

4.2.1 Mechanical Design

SOTAB-II, an autonomous sailboat (Fig. 4.2), is a small unmanned mono hull capable of fully autonomous navigation using global positioning system (GPS) data and attitude sensors based on the desired trajectory generated by its onboard guidance, navigation, and control (GNC) system. It has been designed and developed by Osaka University's Kato laboratory since 2010. A previous study (Senga et al. 2013) dealt with an earlier version of SOTAB-II. Sea experiments were conducted using the early cylindrical hull SOTAB-II at Osaka Bay, about 7 km from Awaji Island in Japan (December 21 to December 23, 2011); and lake experiments were conducted at Lake Biwa (September 29 to September 31, 2011). Three millimeter thick neoprene rubber sheets were used to simulate the oil spill. The experimental results showed that while sail control was achieved, the buoy's drifting velocity was unable to match drifting speed of the target. From this experiment, it was concluded that SOTAB-II needed a larger sail to generate the required force and that the hull shape needed to be redesigned because the drag force created by the cylindrical shape of the hull was excessive, thus making auxiliary propulsion for catching up with the oil spill difficult to achieve.

4.2.1.1 Hull Design

A new yachted-shape SOTAB-II was designed to overcome these problems (Fig. 4.2). This new SOTAB-II is a flexible robotic platform, capable of carrying payloads and sensor equipment and of storing this data onboard. The KIT34 sailing yacht hull form, designed by Kanazawa Institute of Technology, was chosen for the new SOTAB-II hull design (Masuyama et al. 1993) due to its wider beam and larger displacement as compared to other sailing yachts of similar hull dimensions. The KIT34 hull form possesses the space requirements necessary for batteries, motors, data acquisition, and control electronics. Initially, the hull size of SOTAB-II was intended to be one fifth the size of the KIT34 hull (Senga et al. 2013); however,

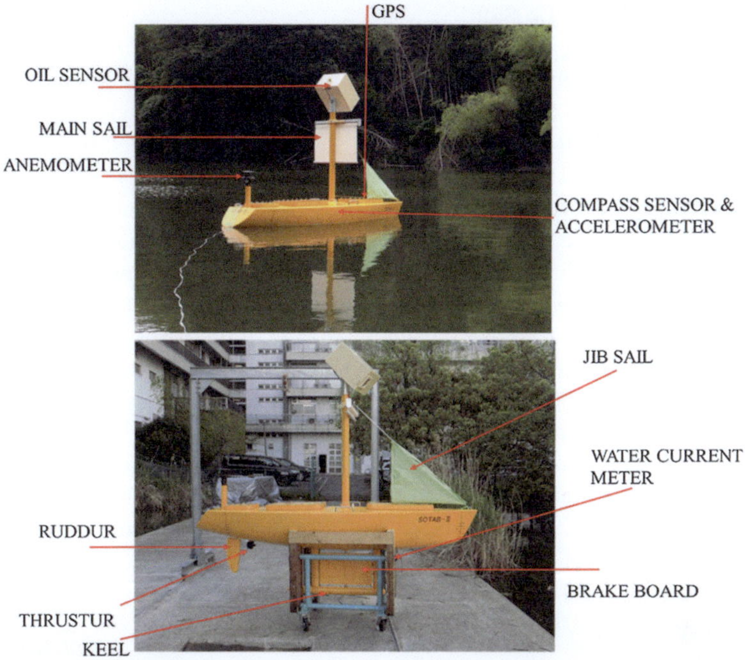

Fig. 4.2 Configuration of SOTAB-II

Table 4.1 Physical dimensions of SOTAB-II

Total length (LOA)	2.64 m
Maximum width (beam)	0.76 m
Draft	0.61 m
Mast height	1.60 m
Displacement	1470 N
Keel position from hull bottom	0.40 m
Keel weight	294 N

the incorporation of the oil sensor (Fig. 4.2), mounted at an altitude of 1.6 m above the deck and weighing approximately 10 kg with all fittings, raised SOTAB-II's center of gravity and reduced the metacentric height (GM). The quarter-sized hull was subsequently settled upon. At $\frac{1}{4}$ the original hull size, the SOTAB-II hull is a scaled-down version of the KIT34 (Rathour et al. 2014). The righting lever was found to be 147.6 mm, with a draft of 146 mm from the hull bottom, and the vanishing point of stability was found to be 125°. The main hull dimensions are indicated in Table 4.1.

4.2.1.2 Mainsail and Jib Sail

SOTAB-II is fitted with two sails: jib sail and mainsail. The jib sail acts as a passive actuator to keep SOTAB-II aligned with the wind. The mainsail captures the wind to provide SOTAB-II with propulsion and modifies its speed by changing the length of the mainsail. The mainsail has only one point of freedom, thus alternating between furling and unfurling. The proportions of the mainsail (0.75×0.75 m) were designed using a computational fluid dynamics (CFD) simulation. In general, the drifting speed of an oil slick is approximately 3.0 % of wind speed along the direction of the wind at a height of 10 m from the sea surface (Unoki 1993). SOTAB-II has been designed to have similar interaction with the waves and surface water currents as that of an oil slick; hence, the core objective of the mainsail is to propel SOTAB-II at a speed consistent with the wind speed affecting the oil spill. The CFD simulations demonstrated that the thrust force provided by the fully unfurled mainsail was greater than the resisting force where the drift velocity of the spilled oil is within 2–4 % of that contributed by the wind velocity. Therefore, the mainsail can provide SOTAB-II with enough thrust until the oil slick, under the condition that the drift velocity of the vehicle is caused by wind, is within 2–4 % of the wind speed (Rathour et al. 2015a, b).

4.2.1.3 Keel Design

As mentioned above, the inclusion of the oil sensor degraded the dynamic stability of SOTAB-II. Therefore, in order to increase stability and prevent the robot from capsizing in rough weather, SOTAB-II was fitted with a 0.4 m keel and a 30 kg ballast bulb, which is 80 mm deeper and 30 kg heavier than what was previously calculated (Senga et al. 2013). Taking down the keel lowers the center of gravitation of the sailboat, which makes SOTAB-II not only more stable but also faster because less driving power is consumed when SOTAB-II remains upright.

4.2.1.4 Brake Board

The other major alteration made to the original design was the addition of a brake board (i.e., a flat plate) to balance the aerodynamic forces acting along the upper piece of the body over the water surface via the hydrodynamic forces acting along the lower section of the body under the water surface. Thus, the brake (i.e., flat plate) can be used to increase roll stability as well as to gain higher speed by keeping the brake in the off position (i.e., longitudinal position). Consequently, a rectangular board, 0.50 m in width and 0.30 m in height, was used for this purpose.

4.2.1.5 Rudder

Unlike the former design of SOTAB-II, the direction of the sail in the new model is fixed. For maneuver control, SOTAB-II is equipped with a single rudder system, the rudder actuators assembled inside the hull are well sealed and protected against water. The rudder of the new SOTAB-II is also controlled by a motor with a movable angle from −90 to 90°, while the effective operating angle is between +35 and −35°.

4.3 Hardware Description

4.3.1 Power Supply

SOTAB-II uses two 12 V lead acid batteries to power the onboard hardware. It also has one DC–DC converter for 12 V and 24 V DC output supply to meet the requirements of the sensor. For powering the onboard computer, SOTAB-II uses a DC–DC converter with a 5 V output. The oil sensor runs on AC power; hence, it has a DC–AC convertor.

4.3.2 Onboard Computer

The hardware components comprising the GNC are located in the center of the hull. The main computer and various peripheral devices, such as serial–USB interfacing hardware, voltage regulator, DC–AC converter, and the wireless LAN hub, are housed in a sealed plastic fiber box. The main information-processing system for the management of data capture and autonomous control of the SOTAB-II hull is an Advantech ARK-1120 L, a 1.66 GHz device with 2 GB RAM and a 256 GB of compact memory (Fig. 4.3a).

4.3.3 Actuators

For actuation of SOTAB-II, rudder angle and mainsail length are trimmed by GNC; the rudder is actuated by a servomotor (CM1-17L30C Cool Muscle with inboard motor controller and encoder) equipped with gear drive. The furling and unfurling of the mainsail are performed by a spring-loaded tendon drive powered by a CM1-17L30C Cool Muscle with inboard motor controller and encoder. One end of the mainsail is fixed to the mast with a fixture mounting by spring. The brake board is driven by a worm and wheel gear drive, which is powered by a servo motor (CM1-23L20C Cool Muscle with inboard motor controller and encoder). The oil sensor shaft is also powered by a servo motor (CM1-17L30C Cool Muscle with inboard motor controller and encoder). A belt drive was used as a power transmission mechanism between the oil sensor shaft and the motor shaft.

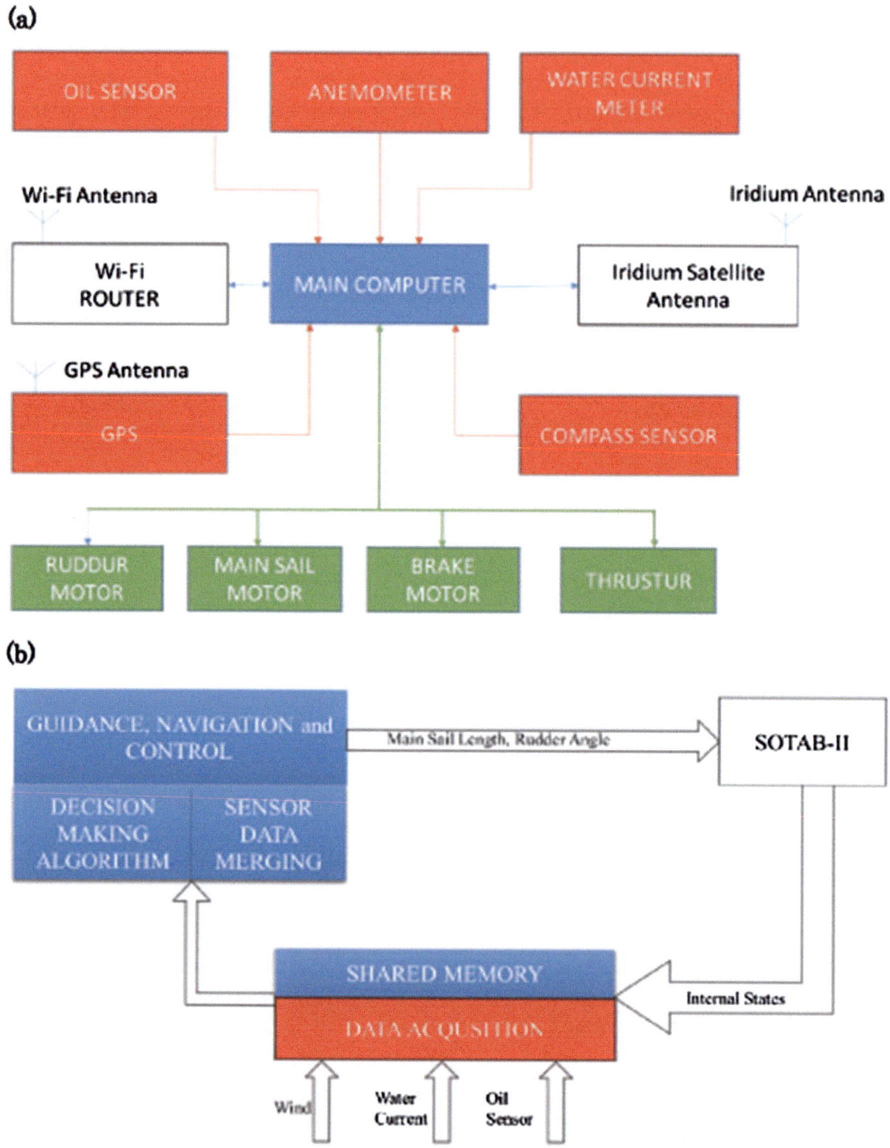

Fig. 4.3 Control system of SOTAB-II: (**a**) hardware architecture, (**b**) software architecture

4.3.4 Sensors

An overview of the hardware and software architecture is illustrated in Fig. 4.2 and described in Fig. 4.3a, b. SOTAB-II uses a number of sensors for its localization and navigational calculation.

4.3.4.1 Guidance and Navigational Sensors

Oil Sensor

SOTAB-II uses a Slick Sleuth SS300 for an oil sensor. The Slick Sleuth SS300 is an optical sensor that can detect micron level amounts of oil in real time from a distance of 1–5 m above the water surface. The Slick Sleuth SS300 comes with a user-selectable detection setting, enabling the user to define the detection threshold and detection period (i.e., sampling interval/frequency), from 0.5 s to 90 min.

GPS

A Hemisphere A325™ GNSS smart antenna is used for obtaining SOTAB-II's position, velocity, and heading. The sensor can output data at a rate of 1, 2, 10, and 20 Hz. In this study, we are using an output data rate of 1 Hz to record the position of the buoy each second.

Compass Sensor

For the azimuth, roll, and pitch angle of SOTAB-II, we are using a TDS01V, a 3D sensor that can measure onboard 3-axis accelerometer, 3-axis geomagnetic, and absolute atmosphere. The 3D sensor has a variable accuracy and range for obtaining roll, pitch, and azimuth angle. It can measure azimuth from 0 to 360° clockwise positive, pitch angle from 0 to 45°, and roll angle from −45 to 45°.

4.3.4.2 Oceanographic Sensors

Wind Anemometer

The anemometer selected for SOTAB-II is ultrasonic wind anemometer manufactured by Gill Instruments: Environmental and Industrial Monitoring Solutions. For wind direction, the measurement is 0–360° if north to south, 90° east to west, 180° south to north, and 270° from west to east. The anemometer's wind speed and direction resolution are 0.01 m/s and 1°, respectively. The accuracies of the wind speed and direction were within ±2 % and ±3°, respectively.

Water Current Meter

For current meter, AEM-RS an electromagnetic current meter produced my Japan ALCC company is used. The current meter gives water velocity and compass readings with which the current velocity and direction can be calculated. The resolutions of the current speed and direction were 0.02 cm/s and 0.010, respectively. The measurement accuracies of the current speed and direction were within $\pm 2\,\%$ and $\pm 2°$, respectively. The results of sensor speed and direction calibration testing are published in Rathour et al. (2015a).

4.3.5 *Communication*

As explained above, SOTAB-II uses various sensors and actuators for its localization and actuation, each with a different communication protocol depending on the firmware implementation, RS-232, CAN, and USB protocols. Manual control of SOTAB-II is also possible via a wireless link, using a remote computer (Fig. 4.3a). For sending data to a land base station, SOTAB-II will use the iridium satellite communication network; although SOTAB-II does not presently have an iridium satellite antenna, it will be added in the near future.

4.4 Software Module

Figure 4.3b shows an overview of SOTAB-II's software architecture. All software was written in Borland C++. This software manages a shared memory and thus provides communication between the individual sensor programs. For example, sensor drivers read the sensor data from the detector hardware and write it to the shared memory. Other programs requiring sensor data for processing can read data directly from the shared storage. The command signals generated by the GNC for sail and rudder control are likewise stored in the shared memory for further analysis (Fig. 4.3b).

4.4.1 *Data Acquisition Module*

Sensor data includes the wind speed and direction, water current and direction, position, azimuth, and relative position of SOTAB-II. These sensors each have their own acquisition frequencies and resolutions. To begin with, the collection and formatting of input data from the sensors are carried out by the onboard computer interfaced by the sensors by the data acquisition module. The data from each sensor is saved in the shared memory with a common clock giving a coherent sense of timing to all the measurements. Secondly, the environmental condition and relative position of SOTAB-II, with respect to the oil slick, is extracted from the sensors and fed to the GNC module.

4.4.2 Guidance, Navigation, and Control Module

The main objective of the GNC module is to reliably detect and localize the oil slick and to supply the control actuators with real-time rudder angle and sail position instructions to achieve the target heading and target speed. Target heading and speed are computed based on the data from the oil sensor and environmental conditions (e.g., wind direction). The target heading is computed by way of a decision-making algorithm, which computes a new target heading each time the data coming from the data acquisition module is refreshed. The target heading is modified to compute the best feasible heading, considering physical constraints of SOTAB-II. The mainsail position is computed based on the target speed and SOTAB-II's speed. The target heading and target speed derivation algorithm is described in detail in Sect. 4.5. The desired rudder angle and mainsail position, coming from the GNC module, is used as a reference for the PID controllers designed for controlling the rudder and mainsail. The PID control algorithm for mainsail and rudder is described in detail in Sect. 4.5.

4.5 Guidance, Navigation, and Control

4.5.1 Coordinate System and Sensor Configuration Description

As described in Sect. 4.2, SOTAB-II uses various sensors for navigation. This section describes the decision-making algorithm for deriving the control input from that sensor information. Altogether, SOTAB-II relies on five sensors for autonomous oil spill detection and tracking. Some sensor data is measured by an absolute value, such as GPS data using the earth coordinate; on the other hand, some data is defined by relative values, such as the wind speed measured by the anemometer, which only provides relative values in body coordinates. Consequently, it is necessary to coordinate between these disparate data forms by merging them into the same form. Coordinate systems are defined for the control system design by XY (i.e., earth-fixed coordinate system) and $X'Y'$ (i.e., body-fixed coordinate system) (Fig. 4.4). In Fig. 4.4, V is speed of SOTAB-II, α is drift direction of SOTAB-II, γ is relative wind direction, ς is water current direction, ψ is direction of resultant vector of water current vector and wand vector, $p(x,y)$ is position of SOTAB-II, and $p(x_k,y_k)$ is position of k-th oil slick. Vector addition of the measured data is carried out for coordinate transformation and a detailed implementation of the same can be found in Rathour et al. (2015a, b).

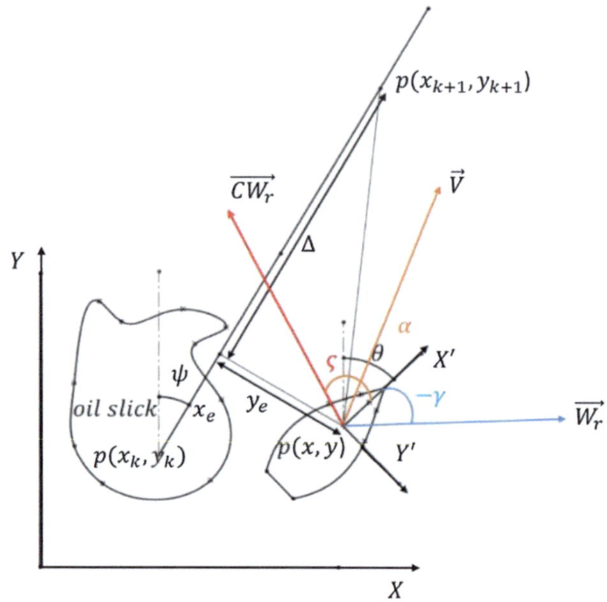

Fig. 4.4 SOTAB-II coordinate system

4.5.2 Decision-Making Algorithm

SOTAB-II uses an optical-based oil detection sensor that detects the fluorescence of oil. The sensor transmits a cone-shaped beam of ultraviolet light at an angle of 14° to the target to be monitored. After absorbing the ultraviolet rays, the target reflects light in the visible spectrum, which is then captured by the receiver. If the captured fluorescence matches the oil fluorescence, the sensor generates a pulse verifying the presence of petroleum. Different oils will have different fluorescence values; therefore, the sensor sets the threshold of the detecting circuit according to user requirements. Based on the data received from the oil sensor, the decision-making algorithm can determine whether the robot is moving away from or drifting with the oil slick. Because oil sensor can be focused on a small area, it was decided to rotate the oil sensor by 360° to cover the area around SOTAB-II. This circular target area is divided into 12 sectors of 30° as shown in Fig. 4.5a. The oil sensor is rotated continuously in a footprint of 30° to skim the area around the SOTAB-II. It takes 2 s for the mast to rotate the oil sensor by 30° and 1 s to take a reading of the target area (Fig. 4.5a). For each scan, the circular position of the oil sensor and its reading is stored in the shared memory. The positional value of the oil sensor is defined in respect to the body axis frame with the absolute angular position shown in Fig. 4.5a. The target heading and target speed of SOTAB-II is decided after determining the robot's present position relative to the oil slick as determined based the on the oil sensor data. Figure 4.6 shows an overview of the decision-making algorithm. Thirteen readings need to be taken around the SOTAB-II to determine whether the robot is within the spill, out of the spill, or at the border of the spill. To calculate

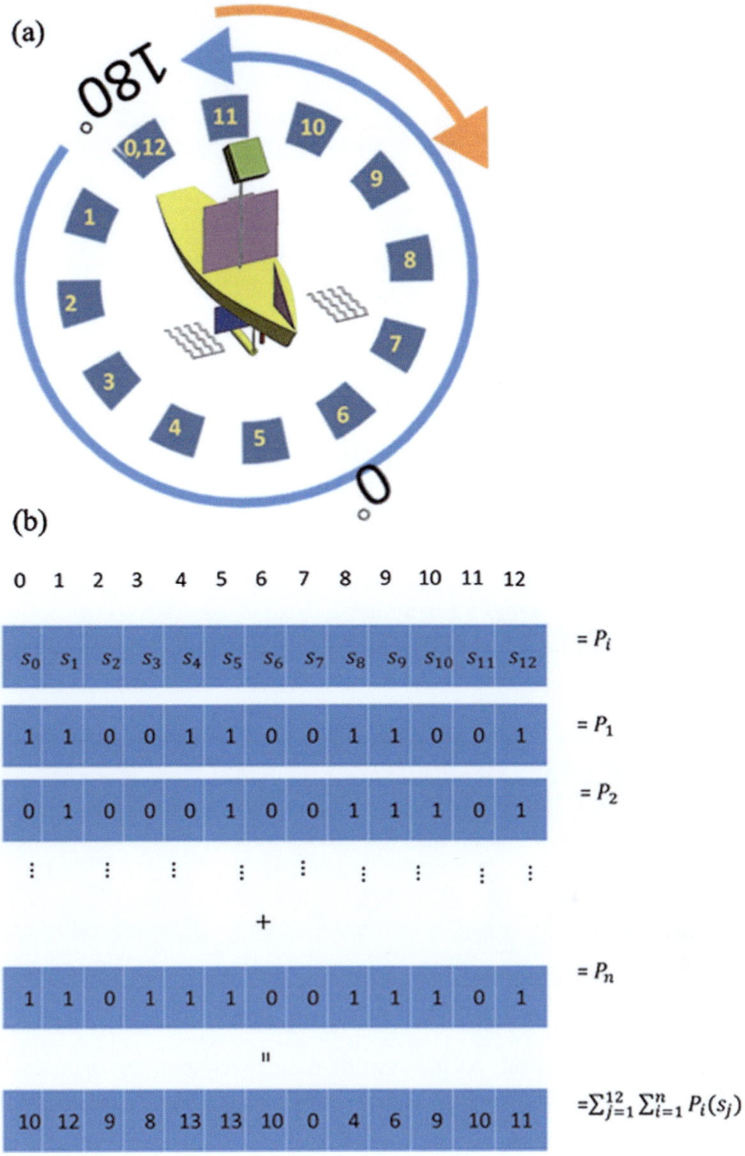

Fig. 4.5 Acquisition of dataset by oil detection sensor: (**a**) target scan area and target points, (**b**) time history sensor dataset and current sensor dataset

the target heading (*TD*) and target speed (*TV*), the maximum number of detected sensor points is computed (Fig. 4.5b). The 13 readings for one complete rotation of the oil sensor are expressed in Eq. 4.1. P_i in (4.1) represents the oil sensor dataset for one rotation of the oil sensor, where 1 denotes successful oil detection and 0

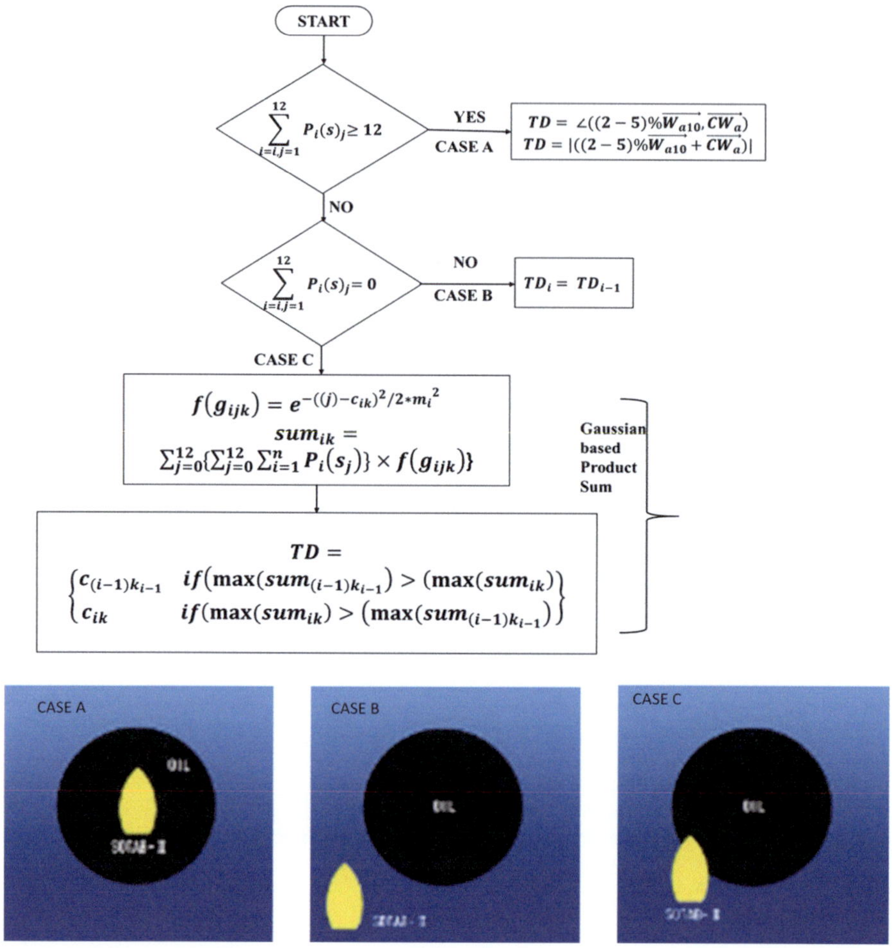

Fig. 4.6 Decision algorithm flowchart

denotes a failure to detect oil. At the start of the experiment, the decision-making algorithm initially waits for the first sensor dataset $P_{i=1}$, after which P_i is updated every 3 s as it takes 3 s for the sensor to move to a new position and compute the target area reading (s_j, Fig. 4.5). Based on the maximum number of detected sensor points in sensor dataset (i.e., $\sum P_i\left(s_j\right)$ in each P_i (Fig. 4.5b)) and the time history of the cumulative sum of sensor data for each of the target points expressed in (1) (i.e., $\sum_{j=1}^{12}\sum_{i=1}^{n} P_i\left(s_j\right)$ (Fig. 4.5b)), the following rules were derived:

$$P_i = \{s_0, s_1,\ s_2, \ldots s_{12}\}, s_j = 1\ or\ 0\ (j = 0, 1,\ 2,\ 3, \ldots 13) \tag{4.1}$$

4.5.2.1 Case A (Target Heading and Target Speed Derivation if SOTAB-II Is Surrounded by Oil)

If all 13 readings expressed in Equation 4.1 are true and oil is detected by the oil sensor (i.e., $\sum s_j \geq 12$), then it can be concluded that SOTAB-II lies within the oil spill (Fig. 4.6). As long as SOTAB-II is within the oil spill, *TD* and *TV* will be calculated as follows:

$$TD = \psi \quad \text{where} \quad \psi = \angle\left((2\text{–}5\%)\,\overrightarrow{w_a}, \overrightarrow{cw_a}\right) \tag{4.2}$$

$$TV = \left|(2 - 5\%)\,\overrightarrow{w_a} + \overrightarrow{cw_a}\right| \tag{4.3}$$

where *Wa* is the wind speed and *CWa* is the water current speed.

4.5.2.2 Case B (Target Heading and Target Speed Derivation if SOTAB-II Is Out of Oil Slick)

If all the 13 readings expressed in Eq. 4.1 are not true and oil is not detected by the oil sensor (i.e., $\sum s_j = 0$), then it can be concluded that SOTAB-II lies out of the oil spill (Fig. 4.6). In such a situation, from the time history of the oil sensor dataset, $p(x_k, y_k)$ can be derived as the best known position of SOTAB-II where oil was. Furthermore, if SOTAB-II is made to converge upon the line defined by the waypoint $p(x_k, y_k)$ where oil was detected and the slope of the line is given by the resultant direction of the water current and wind (ψ), SOTAB-II can be guided back to the oil slick. This is the same as a look ahead-based line-of-sight (LOS) algorithm (Lekkas 2012). The waypoint $p(x_k, y_k)$ is kept on projecting by a distance defined by ($t \times TV$), t is algorithm update time, unless SOTAB-II finds the oil slick. In this case, the *TD* is stated in (4.5). In Eq. 4.4, y_e is a cross track error (Fig. 4.4):

$$y_e = -(x - x_k)\sin(\psi) + (y - y_k)\cos(\psi) \tag{4.4}$$

$$TD_i = \psi + \arctan(-y_e/\Delta) \tag{4.5}$$

4.5.2.3 Case C (Target Heading and Target Speed Derivation if SOTAB-II Is Tracking the Edge of the Oil Slick)

If the number of true sensor readings is less than 12 (i.e., $\sum_{j=1}^{12} P_i\left(s_j\right) < 12$), it implies that SOTAB-II lies on the edge of the spill (Fig. 4.6). Based on the relative position of SOTAB-II with respect to the oil slick, various conditions can be defined.

Clustering Algorithm

To deal with the worst case situation (i.e., where SOTAB-II is found to be surrounded with multiple patches of different sized of slicks), the sensor dataset (P_i) is scanned to determine the total number of large oil slick patches surrounding SOTAB-II (i.e., k). Therefore, the total number of clusters surrounding SOTAB-II is given by k and the center of each cluster can be derived using Eq. 4.7.

Gaussian-Based Oil Slick Cluster Filtering

The center of each slick patch gives the Gaussian function mean (c_{ik}, (4.7)), and the length of the detected target point in each cluster of oil slick patch gives the standard deviation of the Gaussian function (m_i). In (4.7), j_{ik} is the starting indices of oil slick cluster subarray having k_i number of detected points. The Gaussian function dataset formed using (4.6) is multiplied element-wise with the cumulative sum of target points or time history of target points (i.e., $\sum_{j=1}^{12}\sum_{i=1}^{n} P_i\left(s_j\right)$); ((4.8), Fig. 4.7). The total product sum of the element-wise multiplication of $\sum_{j=1}^{12}\sum_{i=1}^{n} P_i\left(s_j\right)$ and Gaussian functions (4.6) was compared to find the largest product sum (sum_{ik}). Target heading (TD) is given by the center of the oil slick patch having largest product sum using (4.9):

$$f\left(g_{ik}\right) = e^{-((u)-c_{ik})^2/2*{m_i}^2} \quad \text{Where}$$
$$u = \begin{cases} j - c_{ik} & \text{if} \quad \left(abs\left(j - c_{ik}\right) \le 6\right. \\ \left(6 + c_{ik}\right) - abs\left(6 - j\right) & \text{if} \quad \left(abs\left(j - c_{ik}\right) > 6\right. \end{cases} \qquad j = 0, 1 \ldots\ldots 12 \tag{4.6}$$

$$c_{ik} = \frac{2 * j_{ik} + m_i - 1}{2}, \; k = 1, 2, \ldots .k_i \tag{4.7}$$

$$sum_{ik} = \sum_{j=0}^{12}\left\{\sum_{j=0}^{12}\sum_{i=1}^{n} P_i\left(s_j\right)\right\} \times f\left(g_{ik}\right) \tag{4.8}$$

$$TD = \begin{cases} c_{(i-1)k_i-1} & \mathit{if}\left(\max\left(sum_{(i-1)k_i-1}\right) > \left(\max\left(sum_{ik}\right)\right.\right. \\ c_{ik} & \mathit{if}\left(\max\left(sum_{ik}\right) > \left(\max\left(sum_{(i-1)k_i-1}\right)\right)\right. \end{cases} \tag{4.9}$$

While calculating the Gaussian functional value, enough care should be taken in calculating the distance of the target point from the oil slick patch center, i.e., as the target points are defined in a circular pattern around SOTAB-II, hence, clockwise and counterclockwise distance should be considered, and minimum of both should be taken. The mathematical average of the angular position of target points within

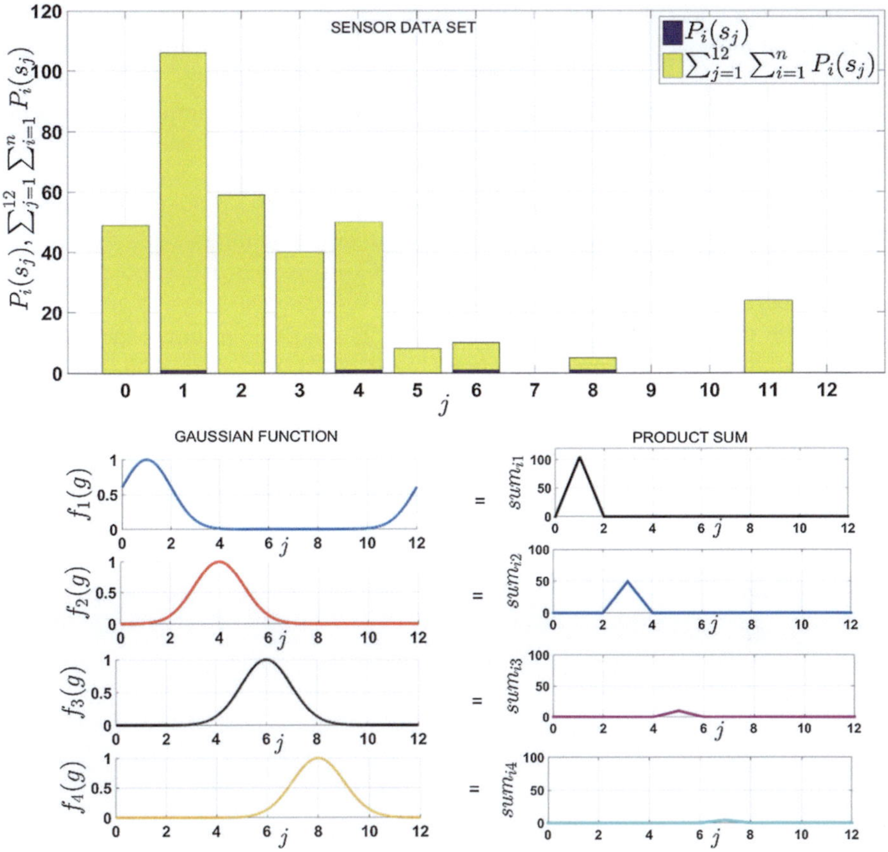

Fig. 4.7 Gaussian-based product sum to derive target heading

the group having largest sum_{ik} will be target heading. Target heading will be given by (4.9).

4.5.3 Guidance and Navigation

Over the years, several methods have been proposed for the path planning of wind-propelled vehicles, the majority of which have been based on fuzzy control theory, neural networks, or other artificial intelligence techniques (Abril et al. 1997; Stelzer et al. 2007; Stelzer and Proll, 2008). Local path planning using the potential method (Petres et al. 2011; Plumet et al. 2014) was also used to navigate the sailboat. However, in all the aforementioned work, none of these studies used a ship's dynamic equation. Other studies, such as Cruz and Alves (2010), used a model with three degrees of freedom for controlling the heading of the sailboat. Nevertheless, the highly nonlinear dynamic model of sailboats, due to complex hull aero- and hydrodynamic models, combined with various appendages (e.g., sail, keel, and rudder), is not always suitable for optimal control of sailing vessels. The use of wind and water current for propulsion of the ASV poses a significant limitation to sailing vessels, as the vessels lose their propulsive force from the environment as they enter “no-go zones” or zones of both upwind and downwind (Plumet et al. 2014; Stelzer et al. 2008). The sailing vessel can get stuck in a no-go zone if it enters too slowly. Following the sailboat behavior described in the aforementioned work, the circular oil sensor target zone was divided into three zones (Fig. 4.8b). The objective is to steer the SOTAB-II toward the center of the slick whenever it tries to go along the edge of the oil slick, avoiding upwind. With this objective in mind, the modified target heading (TD') and modified target speed (TV') were derived based on the region in which target heading lies and the wind heading in body coordinate. For navigation of SOTAB-II, wind heading in body coordinate was converted to $-180° \leq \gamma \leq 180$; with this range, conversion of wind heading, oil senor target points, and wind heading will read negative starboard side and positive port side. TD' and TV' will be determined using the following rules (4.10–4.14):

$$TD' = TD \text{ if } (\ |TD| \leq 75) \vee ((75 < |TD| < 150) \text{ and } ((\gamma \times TD) > 0) \tag{4.10}$$

$$TD' = TD + 180 \text{ if } |TD| \geq 150 \tag{4.11}$$

$$TD' = 180 - \gamma \text{ if } 75 < |TD| < 150 \text{ and } ((\gamma \times TD) < 0) \tag{4.12}$$

$$TV' = TV + 0.1\% \left|(2 - 5\%)\,\overrightarrow{w_a} + \overrightarrow{cw_a}\right| \quad \text{if } |TD| \leq 75 \tag{4.13}$$

$$TV' = TV - 0.1\% \left|(2 - 5\%)\,\overrightarrow{w_a} + \overrightarrow{cw_a}\right| \quad \text{if } |TD| \geq 105 \tag{4.14}$$

4.5.4 Control

SOTAB-II aims to track the oil spill by continuously keeping itself surrounded by the oil that it is tracking. In our control system layout, the rudder and mainsail are modeled as separate single input and single output (SISO) systems and proved to be reasonable during testing (Fig. 4.9).

4.5.4.1 Sail Control

As explained above, the drift velocity of SOTAB-II depends solely upon the mainsail. The sail length is controlled via a PID controller (Fig. 4.9). The control value for the mainsail is its length. The reference value is derived from the

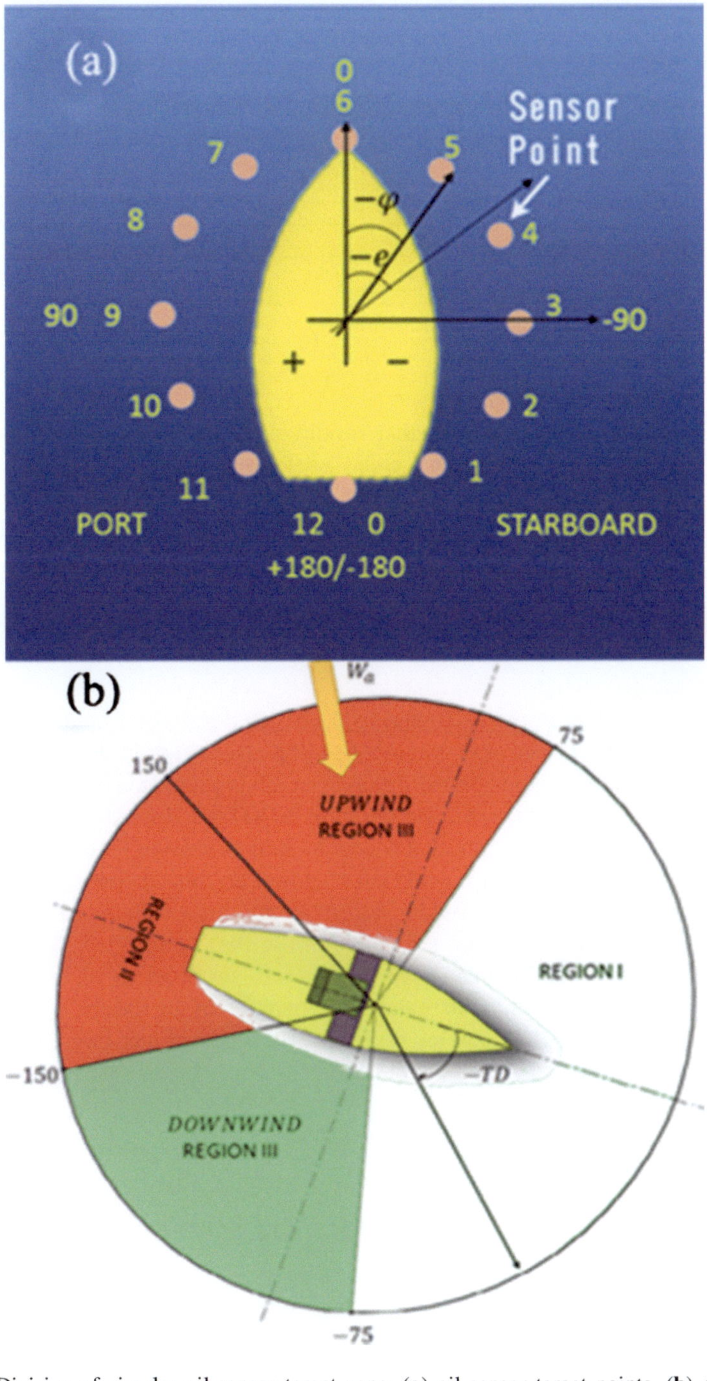

Fig. 4.8 Division of circular oil sensor target zone: (a) oil sensor target points, (b) division of circular zone around SOTAB-II for navigation

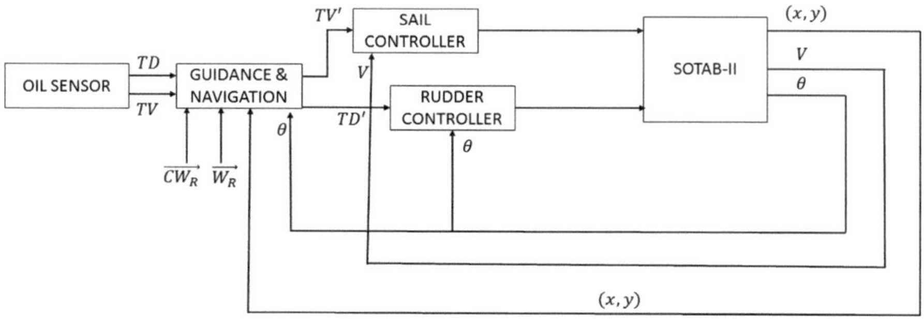

Fig. 4.9 Control flow diagram

perception algorithm, which is further modified by the navigation module. TV', derived from the oil sensor dataset, acts as a reference point for the PID controller. Moreover, SOTAB-II's drifting speed (i.e., V) provides the feedback for the PID loop. If $V > TV'$, then the sail area is reduced, and if $V < TV'$, the sail area is increased.

4.5.4.2 Rudder Control

The maximum rudder angle is limited to $\pm 30°$. Rudder motor control signals are generated by the PID controller where a heading error (e) is in the range of $\pm 20°$; if out of this, a rudder range control signal of $\pm 30°$ is generated based on the e sign (Fig. 4.8 (left)). Therefore, e will be negative if SOTAB-II needs to turn starboard and positive if it needs to turn port. TD' can be in the range of $-165° < TD' < 165°$ (i.e., Region I or Region II) (Fig. 4.8 (left)). However, this design is problematic as higher TD' values may lead to rudder saturation and integral windup. To avoid this problem, TD' is scaled down by a factor g in the PID control law. Therefore, the heading error can be defined as shown in Eq. 4.15. In order to anchor the robot's position to the oil slick within the maneuverable range of the rudder, a heading error input to rudder PID control law for tracking of a time-varying target heading was designed according to Eq. 4.16:

$$e = g * TD' - \theta \tag{4.15}$$

$$\delta(t) = e(t) + Kd\dot{e(t)} + K_i \int e(\tau)\, d\tau \tag{4.16}$$

Broadly speaking, the relative position of SOTAB-II with respect to the slick can be divided into three states: in mode or surrounded by oil (i.e., Case A, $\sum_{j=1}^{12} P_i(s_j) \geq 12$), out of slick (i.e., Case B, $\sum_{j=1}^{12} P_i(s_j) = 0$), and edge mode (i.e., Case C, $\sum_{j=1}^{12} P_i(s_j) < 12$). Consequently, SOTAB-II autonomously maneuvers itself to move inside the slick if $\sum s_j < 12$, by not varying TD' in (15) during edge or out modes unless it has achieved the required maneuver.

4.5.4.3 Brake Board Control

In the case of the brake board, only two positions are selectable: off and on. Off denotes the condition for decreasing the drag force where the face of the brake board with the maximum surface area is oriented along the longitudinal direction of SOTAB-II. On denotes the condition for increasing the drag force where the face of the brake board is perpendicular to the longitudinal direction of SOTAB-II. Respective brake board orientation is determined based on the dynamic responses of SOTAB-II. The brake board, therefore, is positioned off when there is insufficient wind force to provide SOTAB-II with the thrust needed to catch the oil slick or if it loses track of the oil slick and has to look for it again.

4.6 Experimental Results

Various attempts were made to optimize the operation of the control scheme. However, before testing the system in rough water, field experiments were performed in the Osaka University pond (Fig. 4.10a) to ensure that the system was ready for open waters. The free drift and controlled drift pond experimental results have already been published (Rathour et al. 2015a, b). In this paper, the results of a simulated oil spill tracking experiment conducted in Osaka University pond, using a cluster-based decision-making algorithm, is presented (Fig. 4.10c). Free drift, controlled drift, and simulated oil spill (using neoprene sheets) tracking sea experiments, conducted at Eigashima beach, Kobe, are also presented (Fig. 4.10d). The field experiments were conducted to validate SOTAB-II's autonomous oil spill tracking capabilities and to test the guidance and navigation capability of SOTAB-II based on input from the onboard sensors and control logic to derive target headings and directions.

4.6.1 Experimental Results on Free Drift and Controlled Drift

The sail and rudder control experimental results, published in Rathour et al. (2015a, b), demonstrated the effectiveness of the sail and rudder control. However, the sea experiments were carried out on December 3, 2014. During the sea experiment, the wind speed ranged from 6 to 10 m/s, and the water current speed varied from 0.14 to 0.19 m/s. Free drift experiments were performed to characterize the drifting behavior of SOTAB-II. Throughout the free drift experiment, the robot

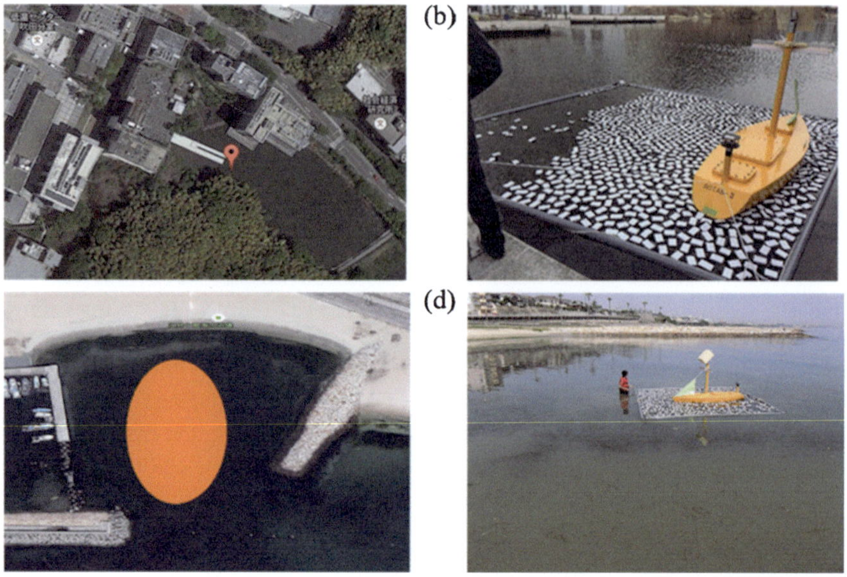

Fig. 4.10 Overview of field experiments of SOTAB-II: (**a**) Osaka University pond experimental site; (**b**) SOTAB-II deployment in the pond with the floating fence; (**c**) Eigashima beach, Kobe sea experimental site; (**d**) SOTAB-II deployment in the sea with the floating fence

was allowed to drift with a fixed sail length and no rudder. Based on a CFD study (Senga et al. 2013), it was established that the flat plate added to SOTAB-II balanced the aerodynamic forces acting on the hull above the water surface and hydrodynamic forces on the hull below the water surface. Hence, following the work of Senga et al. (2013), TD' and TV' for controlled drift were decided as γ and $0.03W_{a10}$, respectively. As mentioned above, $-180° < \gamma \leq 180°$ was considered for rudder control. Significant variations in e can be overruled as the oil sensor was not used in this experiment and $TD' = \gamma$; hence g for this experiment was taken as 1, and e was updated at each interval. The behavior of the system was analyzed offline based on the saved sensor data. This experiment was performed at Eigashima beach, Kobe, Japan, on December 3, 2014. Figure 4.11 shows the trajectory response of SOTAB-II, CWa, 0.03Wa10 and CWa+ 0.03Wa10. It can be concluded from the trajectory response that in controlled mode SOTAB-II was able to follow the resultant trajectory of CW_a and 3 % of W_{a10} very closely; it was found to sail in the resultant direction of water current and wind. The difference in the distance traveled by SOTAB-II and the resultant trajectory of CW_a and 3 % of W_{a10} was found due to slippage in the motor shaft transmission and tendon drive caused by vibration in transportation of SOTAB-II.

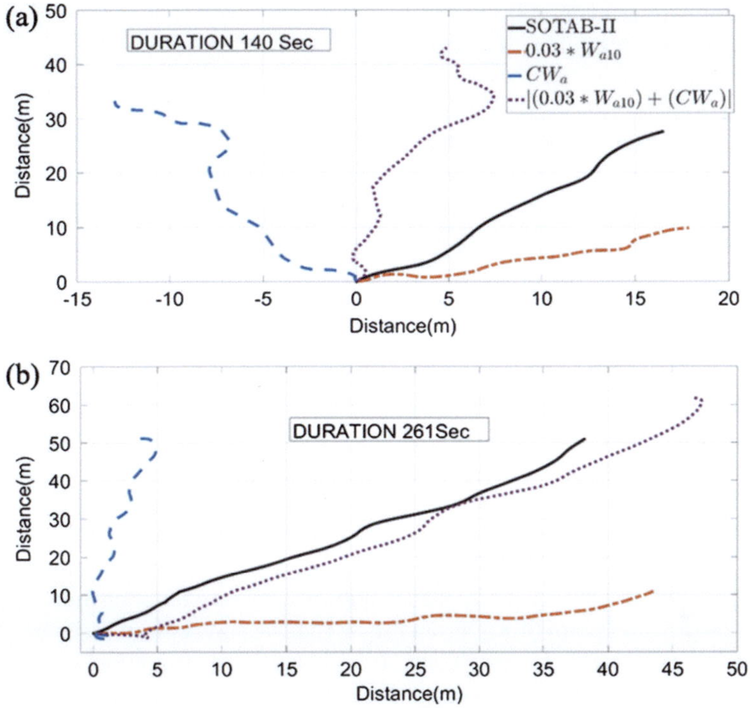

Fig. 4.11 Experimental results at Eigashima beach: (**a**) free drift trajectory, (**b**) controlled drift trajectory

4.6.2 *Simulated Oil Spill Tracking Results*

As we were not permitted to use oil for testing the SOTAB-II control algorithm, we decided to build an artificial oil slick using neoprene sheets. Neoprene sheets were found to drift with a speed similar to that of CW_a and 3 % of W_{a10}. Senga et al. (2013, 2009) and Yoshie et al. (2010) used $1 \times 1 \times 0.03$ m square neoprene sheets for the original SOTAB-II experiments. From the experimental results, it was found that the edges of these neoprene sheets submerged in the water reducing the drift speed. Also, square sheets were found to rotate while drifting to achieve minimum drag; these phenomena made the sheets drift apart. To overcome this limitation, circular sheets were used for our experiment. Neoprene sponge rubber (0.10 m in diameter and 10 mm in thickness), pasted with white Sellotape, was used to simulate an oil spill in this experiment (Matsuzaki and Fujita 2013). This model of target was chosen because the oil sensor can give a false-positive signal when focused on white fabric (Mahr and Chase 2009). A floating fence ($4 \times 4\,m^2$, Fig. 4.10b and d) was used to restrict the initial drift of the neoprene sponge rubber and SOTAB-II. However, the neoprene sponge rubber was found to scatter and rapidly drift outside the detection zone of the oil sensor. Therefore, the control system was evaluated while the oil sensor could still detect them.

4.6.2.1 Experimental Results at Pond

Figure 4.12a shows the time history of m_i (length of largest subarray), k_i (total number of largest subarray), and $\sum_{j=1}^{12} P_i\left(s_j\right)$ (sum of detected points in sensor dataset at each time instant). We can see that at $t = 105$ s, $\sum_{j=1}^{12} P_i\left(s_j\right) = 7$, $m_i = 3$, and $k_i = 2$; this implies that there are two sequences of detected point of each length three. The starting indices of all the sequences have been shown in Fig. 4.12b. From Fig. 4.12b, we can see that the staring indices of two sequences of the largest subarray at $t = 105$ were found to be 0 and 9. This implies "Case C" needs to be followed for derivation of *TD*. Following the steps mentioned in "Case C" of perception, two Gaussian functions with medians of 1.5 and 10 and a standard deviation of 3 were built. The probability of each target point belonging to the three clusters was decided based on (6–7). Fig. 4.12c shows the time history of cumulative sum of each cluster (product sum of Gaussian function and cumulative sum of sensor target point (4.8)). From Fig. 4.12d, we can confirm that cluster 1 will have the largest weighted sum among both clusters, which implies that the target heading will be the center of cluster 1 (i.e., $TD = -105°$). The same was reflected in the time history of *TD* (Fig. 4.12d).

Figure 4.12 (D) shows the time history of the target heading (*TD*), modified target heading (TD'), and relative wind direction ($-180 \leq \gamma \leq 180$) at each instant of time. *TD* is calculated based on (4.9). As explained in the perception section, the time instant sum of every cluster (i.e., max (sum_{ik})) is compared with the previous cluster sum (i.e., max ($sum_{(i-1)k}$)); *TD* is decided based on the cluster having the largest sum value. This was done because there will be a higher probability of the buoy finding the neoprene sheet if it goes in the direction of the cluster with the largest value. This is shown in the time history of *TD* (Fig. 4.12d). From Fig. 4.12c, we can see that cluster with the largest sum was found around time $t = 98\,s$, at the same time instant $TD = -105$.

Figure 4.13 shows extracts from the pond experiment video with their respective frame numbers. These pictures show the relative position of SOTAB-II with respect to the neoprene sheets at different time intervals. A Casio, EXILIM EX-100, operating at 29 frames per second, was used to video the experiment from a nearby rooftop. Frame 1,500 in Fig. 4.13 shows the starting condition of the experiment. SOTAB-II was tied using a rope and a snap knot to the four corners of a fence that could be opened by just pulling the rope from a distance. At the start of the experiment, SOTAB-II found itself surrounded by the sheets, as shown in Fig. 4.13 (F1500); while pulling the fence to allow SOTAB-II and the sheets to drift freely, a few sheets were found to be dragged by the fence and to form a cavity in the bow region of SOTAB-II in Fig. 4.13 (F3500). This phenomena lead to the formation

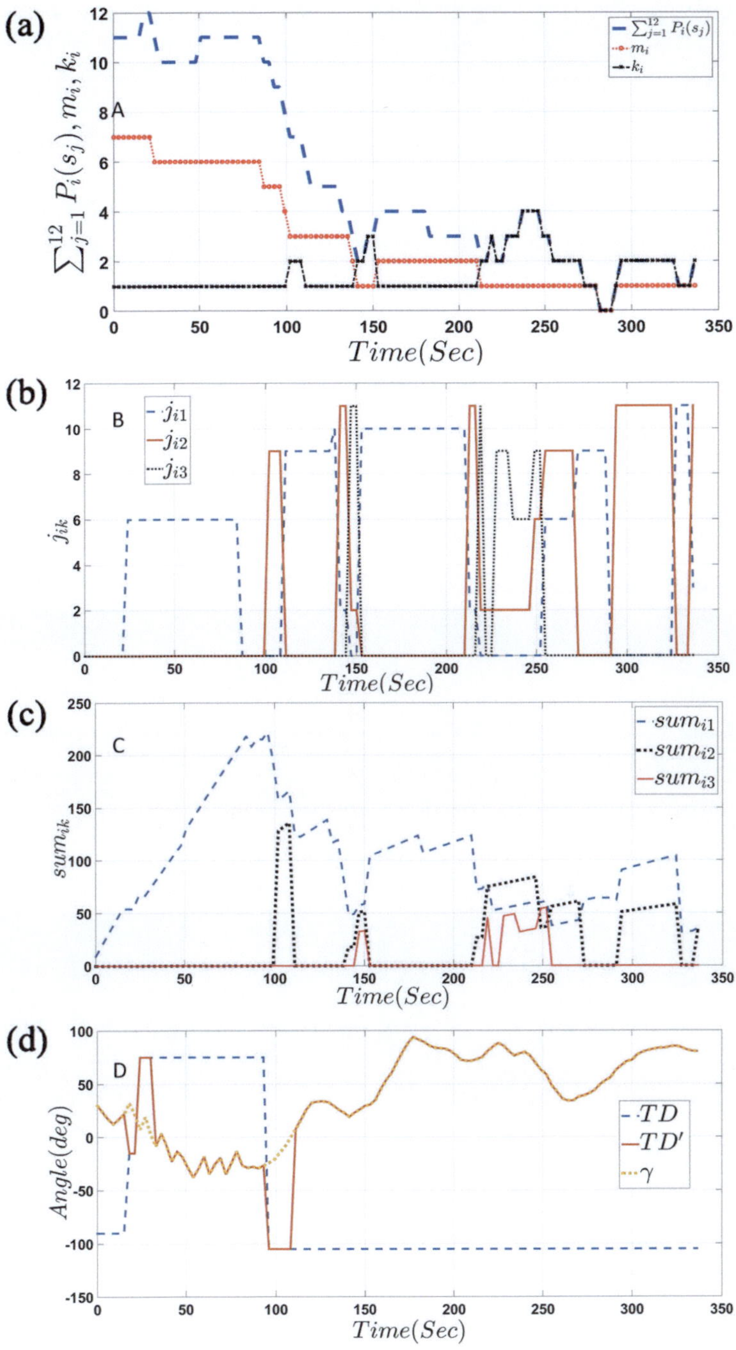

Fig. 4.12 Experimental results at pond: (**a**) total number of target points detected in each scan, length of largest subarray, total number of largest subarray in each sensor dataset; (**b**) starting indices of each largest subarray in sensor dataset; (**c**) Gaussian product sum of each cluster; (**d**) time history variation of *TD*, *TD′*, and γ at every time interval

Fig. 4.13 Pond experiment pictures with respective frame numbers

of two neoprene sheet clusters surrounding SOTAB-II (Fig. 4.12a and Fig. 4.13 (F4500 and F6000)). As explained in the decision-making algorithm, the cluster on the starboard side was found to have the greater Gaussian product sum (Fig. 4.12c, between 100 and 110 s) and $\gamma < 0$; hence the navigation and control module commanded SOTAB-II to turn starboard, as can be seen in Fig. 4.13 (F4500). Following this, the wind direction reversed ($TD \times \gamma < 0$), implying $TD' = \gamma$, this being reflected in the TD' time history (Fig. 4.12d). Following the navigational rule defined above, SOTAB-II tried to follow the neoprene sheet by moving downwind. As shown in Fig. 4.13 (F10500), SOTAB-II was found at the end of the experiment to come to rest within a cluster of neoprene sheets and the largest cluster or patch of neoprene sheets was also found to be drifting in the same direction.

4.6.2.2 Sea Experiment Results

From the results of the controlled drift experiment, it can be established that SOTAB-II drifts with water surface currents; therefore, in order to catch up with the simulate oil spill, the mainsail has to provide enough propulsive force to compensate for the wind-induced drift. Following this assumption, the contribution of CW_a in all the equations mentioned above to formulate *TD*, TD', *TV*, and TV' was neglected. This assumption was found to hold true in the experiments.

The same procedure as described previously for the pond experiment was carried out for sea experiment. Figure 4.14a shows the time history of m_i (length of largest subarray (i.e., oil slick patches surrounding SOTAB-II)), k_i (total number of large subarrays), and $\sum_{j=1}^{12} P_i(s_j)$ (sum of detected points in sensor dataset at each time interval). We can see that at $t = 313\,s$, $\Sigma P_i = 4$, $k = 4$, and $m_i = 1$, implying that there were four clusters of detected points, each of length one. The starting indices of all the sequences are shown in Fig. 4.14b. From Fig. 4.14b, we can see that the staring indices of all the four sequences of the largest subarray at $t = 313\,s$ were found to be 1, 4, 6, and 8. Hence, based on the Gaussian-based product sum algorithm, four Gaussian functions with centers of 1, 4, 6, and 8 and spread of 1 were defined. The probability of each target point belonging to the four oil slick patch groups was decided based on Eqs. 4.6, 4.7 and 4.8. Figure 4.14c shows the time history sum_{ik} of each oil slick patch encountered while tracking the oil slick. From Fig. 4.14c, we can confirm that slick patch 1 was found to have the largest product sum among all the four slick patches, which implies that target heading will be the center of oil slick patch 1 (i.e., $TD = -150°$). This can be seen in the time history of *TD* (Fig. 4.14d).

Figure 4.14d shows the time history of *TD*, TD', and γ at each time interval. *TD* is calculated based on Eq. 4.9. As explained above, the product sum of the Gaussian function dataset and the time history of cumulative sum of sensor data (i.e., max (sum_{ik})) is compared at each time interval with the previous product sum (i.e., max (sum_{ik})), and *TD* is determined based on the oil slick patch with the largest product sum value. This approach was taken because there is a higher probability of the buoy finding the neoprene sheets if it goes in the direction of the oil slick patch with largest product sum value. This was shown in the time history of *TD* (Fig. 4.14d). Comparing Fig. 4.14c, d, it can be shown that the *TD* update rule in Eq. 4.9 followed the time history variation of sum_{ik}.

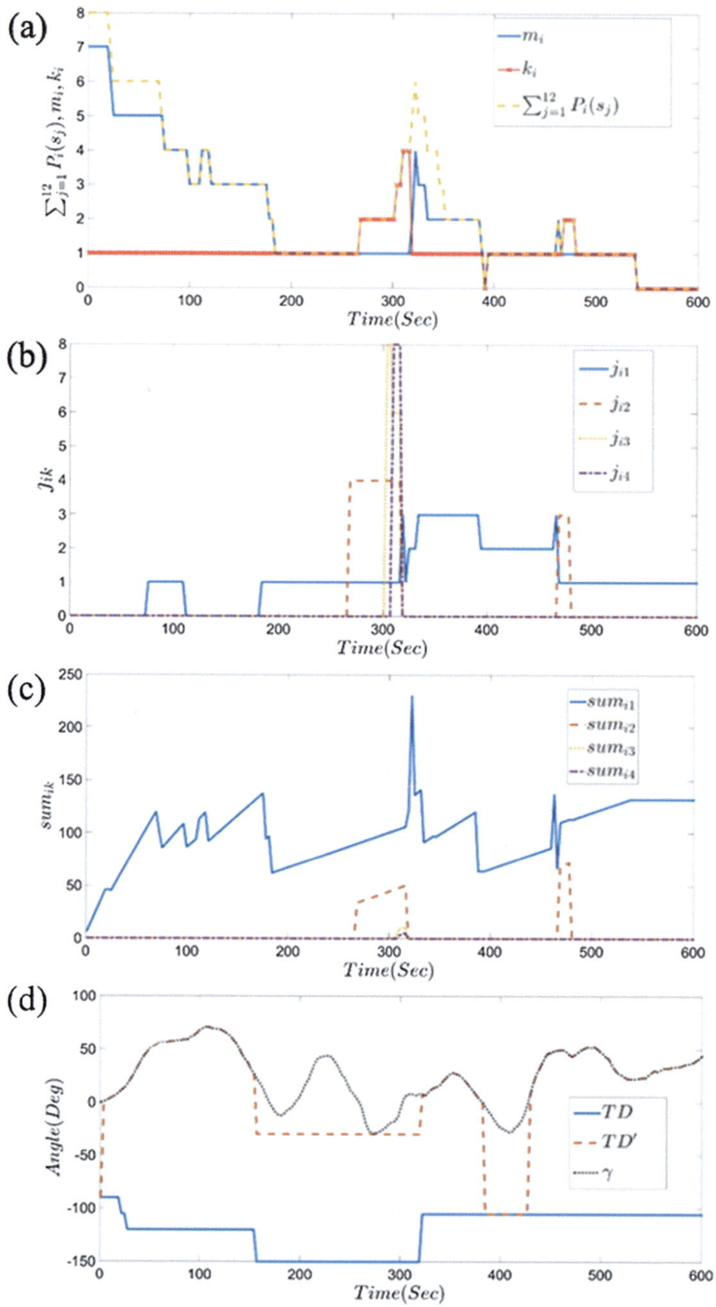

Fig. 4.14 Experimental results at sea: (**a**) total number of target points detected in each scan, length of largest subarray, total number of largest subarray in each sensor dataset, (**b**) starting indices of each largest subarray in sensor dataset; (**c**) Gaussian product sum of each cluster; (**d**) time history variation of *TD*, *TD′*, and γ at every time interval

4.7 Conclusions

This work has described a cluster-based path planning algorithm for the navigation of SOTAB-II. This clustering algorithm, based on the product sum of a Gaussian function and a time history of the cumulative sum of target points, reacts to the changing wind conditions, as well as the dynamic relative positioning of SOTAB-II and simulated oil spills in real time. This method has been successfully tested in both a pond and moderate sea conditions. The sensor-based GNC was validated through field experiments at the Osaka University pond and Eigashima beach, Kobe; the results of which illustrate the free drifting velocity of SOTAB-II, its controlled drift via rudder and sail, and the ability of SOTAB-II to autonomously track simulated oil spills using an oil detection sensor. These experiments have demonstrated that SOTAB-II is capable of moving with a resultant velocity vector at 100 % of water surface currents and at 3 % of wind and that the sail and rudder are effective in controlling the velocity of SOTAB-II within the desired range. Furthermore, the simulated oil spill tracking experiments described in this paper demonstrate that the SOTAB-II was following the largest of the neoprene sheet clusters surrounding the robot, suggesting that the GNC functioned to follow the largest simulated oil spill cluster as anticipated. This cluster-based algorithm can be easily adapted for multiple SOTAB-II deployments.

References

Abril J et al (1997) Fuzzy control of a sailboat. Int J Approx Reason 16:359–375

Alves J, Cruz N (2008) FASt – an autonomous sailing platform for oceanographic missions. In: OCEANS 2008. Institute of Electrical and Electronics Engineers, New York, pp 1–7. doi:10.1109/OCEANS.2008.5152114

Brown CE, Fingas M (2003) Review of the development of laser fluorosensors for oil spill application. Mar Pollut Bull 47:477–484

Cruz NA, Alves JC (2008) Autonomous sailboats: an emerging technology for ocean sampling and surveillance. In: OCEANS Conference. Institute of Electrical and Electronics Engineers, New York, pp 1–6. doi: 10.1109/OCEANS.2008.5152113

Cruz NA, Alves JC (2010) Auto-heading controller for an autonomous sailboat. In: IEEE OCEANS –Sydney, pp 1–6. doi:10.1109/OCEANSSYD.2010.5603882

Fingas M (2011) Buoys and devices for oil spill tracking. In: International oil spill conference. International Oil Spill Conference Proceedings, Washington, DC, p abs9. doi:10.7901/2169-3358-2011-1-9

Fingas M, Brown CE (2007) Oil spill remote sensing: a forensic approach, chapter 14. In: Oil spill environmental forensics: fingerprinting and source identification, Academic Press, Amsterdam

Fingas M, Charles J (2001) The basics of oil spill cleanup, 2nd edn. Lewis Publishers, Boca Raton. 256 p

Goodman R et al (1995) Tracking buoys for oil spills. In: International oil spill conference. International Oil Spill Conference, Washington, DC, pp 3–8. doi:10.7901/2169-3358-1995-1-3, http://www.itopf.com/knowledge-resources/data-statistics/statistics/

Igor I, Lars HR, Lev K, Cecilie W, Johannes R (2012) Comparison of operational oil spill trajectory forecasts with surface drifter trajectories in the Barents Sea. J Geol Geosci 1(105). doi:10.4172/jgg.1000105

Jensen H et al (2008) Recent experience from multiple remote sensing and monitoring to improve oil spill response operations. In: International oil spill conference. International Oil Spill Conference Proceedings, Washington, DC, pp 407–412. doi:10.7901/2169-3358-2008-1-407

Kato N et al (2010) Spilled oil tracking autonomous buoy. In: OCEANS 2010. Institute of Electrical and Electronics Engineers, New York, pp 1–9. doi:10.1109/OCEANS.2010.5664471

Kato N et al (2012) Autonomous spilled oil and gas tracking buoy system and application to marine disaster prevention system. In: Interspill conference. Interspill, London

Lehr W et al (1984) Empirical studies of the spreading of oil spills. Oil Petrochem Pollut 2(1):7–11

Lekkas (2012) A time-varying look ahead distance guidance law for path following manoeuvring and control of marine craft. In 9th IFAC conference on manoeuvring and control of marine craft. pp 398–403. doi:10.3182/20120919-3-IT-2046.00068

Mahr R, Chase CR (2009) Oil spill detection technology for early warning spill prevention. In: OCEANS. Institute of Electrical and Electronics Engineers, New York, pp 1–8

Masuyama Y et al (1993) Dynamic performance of sailing cruiser by full-scale sea tests. In: 11th Chesapeake sailing yacht symposium, Society of Naval Architects and Marine Engineers, Jersey City, pp 161–179

Matsuzaki Y, Fujita I (2013) Horizontal turbulent diffusion at sea surface. J Soc Civil Eng Coast Eng 69(2):460–480

Petres C et al (2011) Modeling and reactive navigation of an autonomous sailboat. In: Proceedings of the IEEE/RSJ International Conference on Intelligent Robots Systems, pp 3571–3576

Plumet F et al (2014) Toward an autonomous sailing boat. IEEE J Ocean Eng 99:1–11

Rathour SS et al (2014) Modelling and control design of spilled oil tracking autonomous buoy. In: 24th international ocean and polar engineering conference. International Society of Ocean and Polar Engineers, Busan, pp 632–640

Rathour SS et al (2015a) Spilled oil autonomous tracking using autonomous sea surface vehicle. Mar Technol Soc J 49(3):102–116 (15)

Rathour SS et al (2015b) Control algorithm for oil spill tracking using ASV with onboard oil detecting sensor. In 25th International ocean and polar engineering conference. pp 645–652 Kona, Big Island, Hawaii, USA, June 21–26: International Society of Ocean and Polar Engineers

Rynne PF, von Ellenrieder KD (2009) Unmanned autonomous sailing: current status and future role in sustained ocean observations. Mar Technol Soc J 43(1):21–30. doi:10.4031/MTSJ.43.1.11

Senga H et al (2009) Spilled oil tracking autonomous buoy system. Adv Robot 23(9):1103–1129. doi:10.1163/156855309X452476

Senga H et al (2012) Verification experiments of sail control effects on tracking oil spill. In: OCEANS 2012. Institute of Electrical and Electronics Engineers, New York, pp 1–7. doi:10.1109/OCEANS-Yeosu.2012.6263416

Senga H et al (2013) Field experiments and new design of a spilled oil tracking autonomous buoy. J Mar Sci Technol 19(1):90–102. doi:10.1007/s00773-013-0233-2

Stelzer R, Proll T (2008) Autonomous sailboat navigation for short course racing. Robot Autonom Syst 56(7):604–614

Stelzer R, Proll T, Robert JI (2007) Fuzzy logic control system for autonomous sailboats. In: 2007 IEEE International fuzzy systems conference, London, pp 1–6. doi:10.1109/FUZZY.2007.4295347

Takagi Y et al (2012) Numerical tracking of methane gas/hydrate and oil droplet in deep water spill. In 11th International conference on global research and education. Inter-Academia, Budapest, pp 1–8

Tsutsukawa S et al (2012) Efficacy evaluation of data assimilation for simulation method of spilled oil drifting. In: Advanced maritime engineering conference. Pan Asian Association of Maritime Engineering Societies, Taipei

Unoki S (1993) Physical oceanography of the coast. Tokai University Press, Tokyo

Yamagishi S et al (2000) Development and test of a compact lidar for detection of oil spills in water. In: Second international Asia-Pacific symposium on remote sensing of the atmosphere, environment, and space. International Society for Optics and Photonics, Sendai, pp 136–144

Yoshie M et al (2009) At-sea trial test of an autonomous buoy which tracks drifting oil and observation of in-situ data tracking drifting markers on the sea for predicting location of the spilled heavy oil. In: Proceedings of the 19th international offshore and polar engineering conference. International Society of Offshore and Polar Engineers, Osaka, pp 437–444

Yoshie M et al (2010) A study on the autonomous buoy system tracking emulsified fuel oil on water through in-situ drifting experiment of the fundamental buoy system model and rubber sheets. J Jpn Soc Naval Archit Ocean Eng 10:189–192

Chapter 5
Numerical Simulation of Oil and Gas Blowout from Seabed in Deep Water

Youhei Takagi, Yasunori Okano, and Naomi Kato

Abstract To simulate oil and gas blowouts from seabed in deep water, we developed two numerical models based on conventional plume and Lagrangian approaches. The simulations were carried out for the Deepspill experiment at the Norwegian Sea, and the two test cases of oil and gas discharges were considered. It was shown that the simple oil-tracking model only with Lagrangian particle tracking could predict the field experiment if the size distribution of oil droplet at the inlet nozzle remains steady during the rising process. The methane gas spill behavior was solved with the hybrid plume/Lagrangian model, and the methane hydrate model was considered in both the near and far fields. Compared with the non-hydration simulation, it was found that the methane hydration has a significant effect on the fate of spilled methane. Furthermore, the parameter determining growth rate in the methane hydrate model plays an important role in obtaining accurate numerical predictions.

Keywords Blowout • Methane gas • Hydrate • Deep water • Numerical simulation

5.1 Introduction

Numerical simulation is an extremely useful tool to make predictions for various physical events, such as the incidents of "oil and gas blowouts" from seabed. Such incidents often occur at deep water because most oil fields are located there in the ocean and may lead to environmental catastrophes during drilling or operations.

Y. Takagi (✉) • Y. Okano
Department of Materials Engineering Science, Osaka University, 1-3 Machikaneyama, Toyonaka, Osaka 560-8531, Japan
e-mail: takagi@cheng.es.osaka-u.ac.jp; okano@cheng.es.osaka-u.ac.jp

N. Kato
Department of Naval Architecture and Ocean Engineering, Osaka University, 2-1 Yamadaoka, Suita, Osaka 565-0871, Japan
e-mail: kato@naoe.eng.osaka-u.ac.jp

© Springer Japan 2017

N. Kato (ed.), *Applications to Marine Disaster Prevention*,
DOI 10.1007/978-4-431-55991-7_5

Thus, oil companies and researchers have been developing numerical models to simulate possible oil and gas blowouts.

During the initial stage of a blowout near the seabed, the released oil/gas mixture due to breakage forms a large plume structure with an upward flow. The development of such a plume structure, which is a phenomenon that occurs when a fluid is injected into other at a high speed, has attracted great interest and led to a large body of studies conducted by fluid dynamics researchers. For instance, McDougall studied the bubble plume flow in stratified water (McDougall 1978), in which the developed double-plume model indicated that the stratified plume layer is distributed in the horizontal direction at a certain sea level and the end of the branched plume do not reach to the sea surface. Indeed, as McDougall suggested, such a horizontal plume structure in a pure stratification without ambient current flow was actually observed in the experiments (e.g., Seol et al. 2009).

The plume structures forming in deep water due to momentum transfer of the ejected oil/gas mixture change their structural forms when the rising oil droplets and gas bubbles reach at a certain sea level where the average density of the plume structure is balanced with that of the ambient water. This sea level is called neutral buoyancy level (NBL). After reaching the NBL, the plume structure breaks, and the particles of oil and gas separately rise and drift as imposed by the sea current.

These changes developing in two stages in a blowout have been studied by various researchers. For instance, a hybrid blowout model in deep water was suggested by Zheng and Yapa (2003). When the spill model shown in Fig. 5.1 is employed, the lower jet/plume region is called "near field," whereas the upper region where advection and diffusion due to current flow are dominant is called "far field." The critical level between the near and far fields corresponds to the NBL, and the average density of plume core region is used to determine this level. Because the

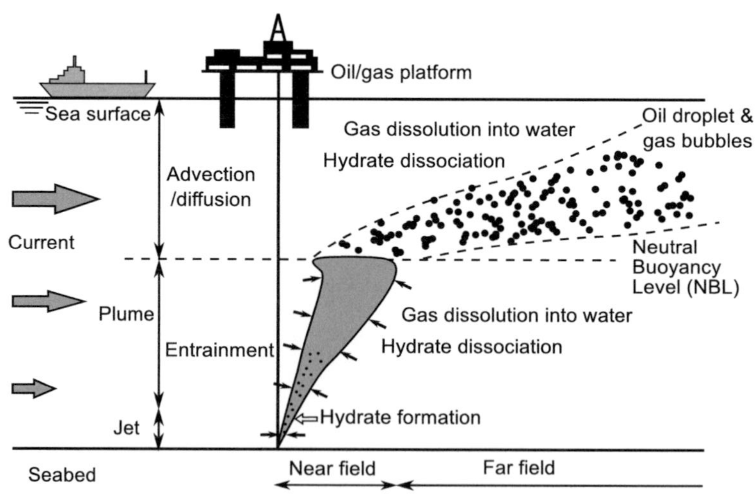

Fig. 5.1 A schematic description of an oil/gas blowout in deep water (Zheng and Yapa 2003)

scale of blowout varies from small bubbles and droplets (less than centimeter) to large plume and slick (more than a few hundreds meters), a hybrid model including some stages of the blowout may be useful for predicting the spill behavior faster and at a low computational cost.

However, the blowout in deep water is more complex and involves various physical phenomena and chemical effects; the model developer of blowouts must consider a model with more options for more accurate predictions. For example, when the blowout of natural methane gas is simulated, we must consider the phase change under the ambient condition of deep water (low temperature and high pressure) and the dissolution of methane gas into the ambient water. Regarding this gas modeling, we discuss it in the following section of this chapter. Recently, the effect of dispersant for segmentalizing oil droplet has been an interest. For instance, the improvement of spill model considering dispersant can be found in the work of Socolofsky et al. (2015).

Various spill models have been developed by research groups and companies, as summarized in Table 5.1. The "plume" tracking model is used for the near field, and the plume structure is discretized with sliced segments along the centerline of plume. On the other hand, the "Lagrangian" tracking model is used for both the near and far fields, and each particle or small parcel including the same particles is individually tracked by using the Lagrangian approach. The DeepBlow (Johansen 2000) is developed by the SINTEF and is based on a simple plume model. The Oil Spill Contingency and Response (OSCAR) (Reed et al. 1999) is a comprehensive spill model including the DeepBlow model as the kernel of the tracking model and can model oil weathering near the sea surface and also the strategic response to accidents. The comprehensive deepwater oil and gas (CDOG) blowout model (Chen and Yapa 2003; Zheng and Yapa 2003) is a hybrid model with the plume model for the near field. The Lagrangian model for the far field was developed by the research group of Yapa at the Clarkson University. The MEthane GAs in DEEPwater (MEGADEEP) model (Yapa et al. 2008, 2010) was also developed by

Table 5.1 Summary of oil/gas spill model

Name	Tracking model	Optional models	Institute
DeepBlow	Plume		SINTEF
OSCAR	Plume	Oil weathering, strategic response	SINTEF
CDOG	Plume, Lagrangian		Clarkson Univ.
MEGADEEP	Plume, Lagrangian	Gas/hydrate	Clarkson Univ.
ADMS/CDOG	Plume, Lagrangian	Bubble size distribution	Clarkson Univ.
LTRANS	Lagrangian		Univ. of Maryland
BLOSOM	Plume, transport, conversion	Weathering, gas/hydrate	NETL
MIKE	Plume, Lagrangian/Eulerian		DHI
OILMAPDEEP	Plume, Lagrangian	Surface weathering	RPS ASA

the Clarkson University with the same tracking model, and this model is specialized in methane gas tracking with hydration. The ADMS/CDOG model (Yapa et al. 2012) is an advanced model of the CDOG, in which the model of bubble size distribution was added and the advection and diffusion modeling for the far field was improved. The Larval TRANSport Lagrangian (LTRANS) model (North et al. 2011) was originally developed in the University of Maryland for oyster larval as the name implies, but it can predict oil spill behavior by using the Lagrangian approach. The BLOwout and Spill Occurrence Model (BLOSOM) is an integrated system to simulate offshore oil spill developed by the National Energy Technology Laboratory (NETL) and includes three main modules for spill tracking: plume, transport, and conversion models. This model can deal with weathering and gas hydration as same as OSCAR and MEGADEEP. The MIKE and OILMAPDEEP are the models included in commercial software and were developed by DHI and RPS ASA, respectively. This software simulates not only oil and gas spill in deep water but also hydrostatic phenomena in shallow water such as river, and the spill model is one of the functions included in the large system. Although the accuracy of model for oil and gas spill in deep water is dependent on the physical and chemical models adopted as the kernel modules of tracking calculations, the prediction of ambient flow conditions also significantly affect the total accuracy of simulated spill behavior. Some models can easily be called a precise ocean model such as ROMS, but the validity of each of these ocean models may be a different issue for spill modeling.

In the present chapter, we discuss two numerical models for spill: (i) simple Lagrangian tracking for oil droplet and (ii) hybrid plume/Lagrangian tracking for gas bubbles. Then, in order to discuss the accuracy and capability of the spill models, the numerical results for the field experiment of "DeepSpill" (Johansen et al. 2001) are shown. The numerical models discussed here are limited with the conventional plume and Lagrangian models. For other numerical modeling techniques such as CFD, the reader is referred to Chap. 6.

5.2 Numerical Model

5.2.1 Simple Oil-Tracking Model

Released components in a spill accident involve many compounds of hydrocarbons. The main components are roughly classified into two groups: crude oil and natural gas. Crude oil is segmentalized, and its droplet size distribution is not significantly affected by the depth level. Therefore, if the chemical effects such as dispersant will not be included in the model, the motion of rising oil droplets can be modeled by the momentum equation alone.

Now, we consider small oil droplets with the diameter distribution varying from 1 to 10 mm. We assume that each droplet does not accompany breakup and coalescence during rising. For simplicity, we resolve the three-dimensional motion of oil droplet into a rising motion in the vertical direction and drifting in the horizontal plane.

As for the rising motion in the vertical (z-) direction, the governing equation is the following momentum balance:

$$(m_o + m_a)\frac{d^2z}{dt^2} = \frac{\rho(S,T,p)}{\rho_o}m_o g - m_o g - F_D \tag{5.1}$$

where z is the coordinate in the vertical direction, t time, m_o the mass of oil droplet, m_a the additional mass acted on oil droplet, ρ the density of seawater, ρ_o the density of oil, g the gravity acceleration, and F_D the drag force. The density of seawater is calculated at each depth level with the equation of state including salinity S, temperature T, and water pressure p (UNESCO 1981). The drag force term is calculated by

$$F_D = \frac{1}{2}\rho(S,T,p)v^2\pi A_o C_D \tag{5.2}$$

where v is the relative velocity of oil droplet to ambient water, A_o the characteristic area of oil droplet, and C_D the drag coefficient. The drag coefficient C_D is dependent on the Reynolds number based on the relative velocity of oil droplet. The empirical relation between the Reynolds number and C_D can be found in the book of Clift et al. (1978), and we selected the relation equations for spherical and elliptical droplets. The governing equation in Eq. 5.1 can easily be discretized by the finite difference method and be temporally integrated by a general explicit scheme such as Runge-Kutta method.

The computation for predicting the horizontal motion due to ambient flow demands significant computing. The conventional approach, that is, the random walk model, is often used to represent the diffusive momentum transfer by unsteady current flow. In this method, the diffusion drifting velocity is given by

$$\vec{U_r} = \vec{\gamma_x}\sqrt{\frac{6D_x}{\Delta t}} + \vec{\gamma_y}\sqrt{\frac{6D_y}{\Delta t}} \tag{5.3}$$

where $\vec{U_r}$ is the diffusion drifting velocity; D_x and D_y are the diffusion coefficient in the $x-$ and $y-$ directions, respectively; γ_x and γ_y are random vectors with the magnitude of $(-1,\ 1)$; and Δt is the increment of time advancement. In order to compute the three-dimensional velocity of oil droplet at each time step, the calculated vertical and horizontal velocities obtained from Eqs. 5.1 and 5.3 are linearly combined. Therefore, this simple model for an individual oil droplet can predict the rising motion of oil affected by the horizontal current flow. In general, since the density of crude oil is smaller than that of the ambient seawater, all released oil droplets reach to the sea surface, and the rising process is practically estimated by the simple tracking model. However, the drifting motion of oil slick or ball along sea surface after the rising process requires other physical models such as the effect of wind waves.

5.2.2 *Hybrid Plume/Lagrangian Model*

As shown in the introduction section (see Fig. 5.1), modeling of oil and gas spill is often considered in two stages: (i) jet/plume model in the near field and (ii) advection/diffusion model in the far field. This hybrid model is useful for simulating a multi-physics phenomenon such as methane gas blowout with phase change. In this section, we briefly review the hybrid spill model for methane gas tracking from seabed. Details of this model can be found in the articles of Yapa (Bandra and Yapa 2011; Chen and Yapa 2004; Zheng and Yapa 2002, 2003).

5.2.2.1 Plume Model: Lagrangian Control Volume Method

Plume structure is stable in the near field around the blowout point, and the global modeling for a large structure is valid for a practical simulation. The plume structure is of a conical shape originated from the blowout point, and its centerline is curved under the effect of strong current flow. The Lagrangian control volume (Lagrangian CV) approach is a reasonable method for capturing this gradually expending structure with low computational demand. The plume structure is discretized with cylindrical segments, that is, with the control volume (CV) shown in Fig. 5.2.

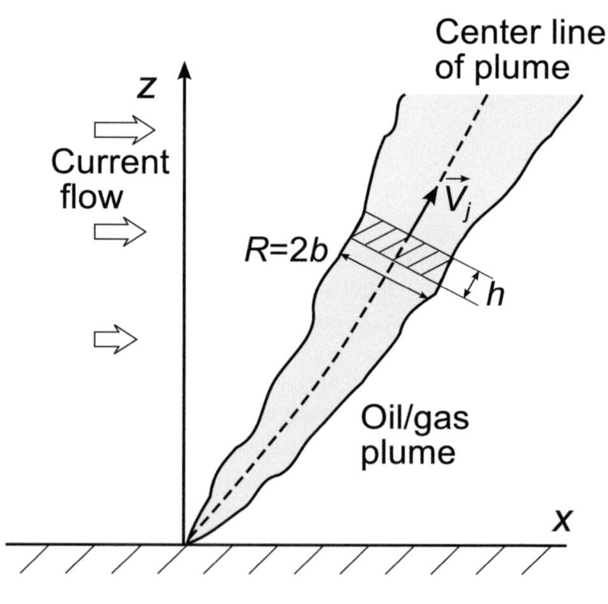

Fig. 5.2 Schematics of the Lagrangian control volume method, where h is the control volume height, b the control volume radius, and V_j the rising velocity

The axis of cylindrical CV corresponds to the centerline of the plume, and the geometrical parameters of each CV are the segment height h and the diameter of circular cross section R. Then, the segment height is determined by the jet velocity and the time increment of computation as:

$$h = |\overrightarrow{V_j}|\Delta t \tag{5.4}$$

A CV moves along the centerline of plume with the jet velocity $\overrightarrow{V_j}$ and is tracked in a Lagrangian framework. Thus, the center positions of CVs are recorded at each time step.

In each CV, the conservation of the fluxes of mass, momentum, and heat must be solved. Additionally, the gas volume (fraction) and bubble size distribution must be also considered in methane gas tracking. In the case of a strong cross flow acting on the plume structure, we need to consider the gas separation from the plume core region and remove the separated bubbles from the CV. Then, the number of gas bubbles included in the CV is calculated as follows (Chen and Yapa 2004):

$$N = \frac{f \cdot J \cdot h}{|\overrightarrow{V_j}| + w_b \sin\psi} \tag{5.5}$$

where f is the remaining gas fraction after separation, J the bubble volume flux into CV, ψ the tilting angle of the centerline of plume from the horizontal plane, and w_b the slip velocity in the vertical direction. If the direction of jet velocity is aligned with that of the slip velocity (straight plume without bending), $f = \sin\psi = 1$, that is, gas bubbles are not separated from the plume core.

5.2.2.2 Conservation Equation in Plume Structure

Under deepwater conditions (high pressure and low temperature), methane may also exist as hydrate (the solid phase). Thus, a model must account for these two phases of methane. The conservation of liquid mass in CV is given by

$$\frac{dm_l}{dt} = \rho_a Q_e - f \cdot J \cdot \tau \cdot n_h \frac{dn}{dt} M_w \tag{5.6}$$

where m_l is the liquid mass in CV, ρ_a the density of ambient fluid, Q_e the entrainment velocity of ambient fluid into CV, τ the traveling time of bubble across CV, n_h the hydrate number of methane, dn/dt the growth rate of hydrate, and M_w the molecular weight of water. The kinetic model of hydrate growth is discussed in the following section.

The loss of methane mass due to hydrate formation and gas dissolution into water is calculated by the following equation:

$$\Delta m_b = -f \cdot J \cdot \tau \left(\frac{dn}{dt} + \frac{dn_s}{dt} \right) M_g \Delta t \tag{5.7}$$

where Δm_b is the loss of methane gas mass in CV, m_b the methane gas mass in CV, dn_s/dt the dissolution rate of methane gas into water, and M_g the gas molecular weight.

The momentum conservation in CV is resolved into three components, the x- and y-components in the horizontal plane and the z- component in the vertical direction:

$$\frac{d}{dt}\left[(m_l + m_b + m_h) \cdot u\right] = u_a \rho_a Q_e - u \rho_{com} Q_g \tag{5.8}$$

$$\frac{d}{dt}\left[(m_l + m_b + m_h) \cdot v\right] = v_a \rho_a Q_e - v \rho_{com} Q_g \tag{5.9}$$

$$\frac{d}{dt}\left[m_l w + (m_b + m_h)(w + w_b)\right] = w_a \rho_a Q_e - w \rho_{com} Q_g + (\rho_a - \rho_l)\, g \pi b^2 (1 - \beta \epsilon)\, h + (\rho_a - \rho_{com})\, g \pi b^2 \beta \epsilon h \tag{5.10}$$

where m_h is the methane hydrate mass in CV; u, v, and w are the average CV velocity component in the $x-$, $y-$, and $z-$directions, respectively; u_a, v_a, and w_a are the ambient flow velocity component in the each direction, respectively; ρ_{com} is the density of hydrate shell including gas inside the shell; ρ_l is the density of liquid in CV; Q_g is the gas volume flowing out from CV; β is the ratio of gas-occupied sectional area to total one of CV; and ϵ is the gas fraction enclosed in hydrate shell to CV. On the right-hand side of Eqs. 5.8, 5.9 and 5.10, the first terms represent the momentum transfer of liquid mass and the second terms stand for momentum loss to flowing gas across the plume side boundary. The other terms including the gravity acceleration on the right-hand side of Eq. 5.10 represent the buoyant forces due to the density difference of liquid and gas with respect to the ambient water.

The energy balance in CV is given by

$$\frac{d}{dt}\left[\left(C_{pl} m_l + C_{ph} m_h\right) T\right] = C_{pl} T_a \rho_a Q_e + f \cdot J \cdot \tau \cdot \frac{dn}{dt} \lambda \tag{5.11}$$

where C_{pl} is the specific heat of liquid at constant pressure, C_{ph} the specific heat of hydrate at constant pressure, T the temperature of plume, T the temperature of ambient water, and λ the latent heat of hydrate formation/dissociation. In the right-hand side of Eq. 5.11, the first term is due to the contribution of entrained ambient water and the second term represents the effect of methane phase change on energy gain/loss in CV.

The conservations of salinity and oil mass are simply given by

$$\frac{d\,(m_l I)}{dt} = I_a \frac{dm_l}{dt} \tag{5.12}$$

where the symbol I means the concentration of salinity or oil mass in CV and I_a is the corresponding component in ambient water.

5.2.2.3 Kinetics of Methane Hydrate

Methane hydration is an important factor in predicting the spilled methane behavior in deep water. At the depth level lower than −600 m, the temperature and pressure conditions of seawater satisfy the condition for forming methane hydrate. Theoretically, the stability condition of methane hydrate is determined by the equilibrium curve of methane, that is, the thermophysical stability. Actually, a rising methane hydrate is directly observed in deep water such as that of Japan Sea (Matsumoto et al. 2009). However, Johansen et al. (2001) reported that hydrated methane was not observed in the field blowout experiment of "Deepspill" in the Norwegian Sea. Although the discussion whether or not methane hydration occurs under actual field conditions requires more detailed studies. We introduce a simple hydrate model in this chapter.

In order to consider the effect of methane hydrate in the spill model, we must take into account the following three physical processes: hydrate growth/dissociation, mass transfer, and heat transfer. One of the well-known structures of methane hydrate is the spherical shell model shown in Fig. 5.3. This shell structure is actually formed when a single methane bubble is injected into the water of low temperature and high pressure in a laboratory experiment.

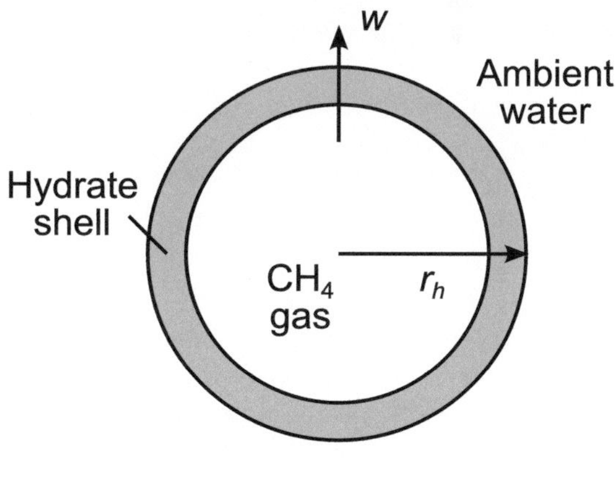

Fig. 5.3 Spherical shell model of methane hydrate in water

When this model is used for simulating the methane gas spill, we make the following assumptions:

1. Enclosed methane gas is diffused through a porous hydrate shell due to the concentration gradient in the radial direction, and methane is hydrated at the interface between the hydrate shell and ambient water.
2. Temperature of the hydrate shell is equal to that of the hydrate-water interface because the heat conductivity of methane gas is much smaller than that of the liquid phase. Thus, the generated heat due to hydrate formation is mainly conducted through the liquid phase.
3. Mass and heat transfer equations are quasi-steady because the movement of hydrate/bubble is very slow under the numerical condition of small time increment.
4. Gas pressure in the hydrate shell is equal to that at the outer shell boundary because hydrate shell has porous structure without pressure difference.
5. Scratching of hydrate is not considered, and hydrate converse the shell surface uniformly.

Under these assumptions, we consider the hydrate formation/dissociation process. The growth rate of methane hydrate is given by

$$\frac{dn}{dt} = K_f A(f_{\mathrm{dis}} - f_{\mathrm{eq}}) \tag{5.13}$$

where dn/dt is the hydrate growth rate, K_f is the growth rate constant, A is the surface area of spherical shell, and f_{dis} and f_{eq} are the fugacities of methane gas under dissolved and equilibrium states, respectively. If the shell shape is a pure spherical shell, the surface area is calculated by

$$A = 4\pi r_h^2 \tag{5.14}$$

where r_h is the radius of hydrate shell. In actual, the shell is porous media, and so we have to correct the original equation of (5.13). Chen and Yapa (2004) introduced an empirical parameter Ψ and modified the growth rate calculation of Eq. 5.13 as follows:

$$\frac{dn}{dt} = K_f \cdot 4\pi r_h^2 \cdot \Psi \cdot (f_{\mathrm{dis}} - f_{\mathrm{eq}}) \tag{5.15}$$

Setting of the empirical parameter Ψ depends on the field conditions and the interpretation of the researcher involved. For instance, Chen and Yapa (2004) suggested the optimized value of $\Psi = 38$ based on the laboratory experiment of Englezos et al. (1987). To improve the dependency of Ψ from the growth equation of (5.15), we have to introduce other coupling equation based on physical grounds.

One of the candidates is the following Ranz-Marshall equation (Ranz and Marshall 1952):

$$Sh = \frac{kr_b}{D_g} = 2 + 0.6Sc^{1/3}Re^{1/2} \tag{5.16}$$

where Sh is the Sherwood number, k the mass transfer coefficient of methane gas, r_b the radius of gas bubble, D_g the effective diffusion coefficient of methane gas, Sc the Schmidt number, and Re the Reynolds number in rising motion. From this equation, we can calculate the value of k and estimate the coefficient K_f considering the convective mass transfer around the hydrate shell. The validity of the usage of the Ranz-Marshall equation is discussed in the following section with a test case, and the results are compared with the parametric calculations for Ψ.

For the simple spherical hydrate shell model, the mass transfer equation can be easily solved under the assumption of quasi-steady state. That is, the concentration profile of methane is calculated by the following boundary value problem from the diffusion equation.

$$\frac{d}{dr}\left(r^2\frac{dC}{dt}\right) \quad \text{for} \quad r_b \le r \le r_h \tag{5.17}$$

B.C.:

$$C|_{r=r_b} = C_0 \tag{5.18}$$

$$C|_{r=r_h} = C_i \tag{5.19}$$

$$-D_g \cdot 4\pi r_h^2 \Psi \left.\frac{dC}{dr}\right|_{r=r_h} = \frac{dn}{dt} \tag{5.20}$$

where C is the concentration of methane gas, C_i the concentration at the hydrate-water interface, and C_0 the concentration at the hydrate-gas interface.

The heat transfer around the hydrate shell is also solved using a similar differential equation:

$$\frac{d}{dr}\left(r^2\frac{dT}{dr}\right) = 0 \quad r \ge r_h \tag{5.21}$$

B.C.:

$$T|_{r=r_h} = T_i \tag{5.22}$$

$$T|_{r=\infty} = T_\infty \tag{5.23}$$

$$-K_w \cdot 4\pi r_h^2 \Psi \left.\frac{dT}{dr}\right|_{r=r_h} = \lambda\frac{dn}{dt} \tag{5.24}$$

where T_i is the temperature at the hydrate-water interface, T_∞ the temperature before hydration, and K_w the heat conductivity of water.

The bubble size of methane gas is calculated by the following equation of state for real gas:

$$P_\infty \cdot \frac{4}{3}\pi r_b^3 = nZRT_\infty \tag{5.25}$$

where P_∞ is the pressure in static water, n the mole number of gas, Z the compressibility, and R the gas constant.

As the kinetics of hydrate dissociation, we can easily model by using the formation equation of (5.13), that is,

$$-\frac{dn}{dt} = K_f A\left(f_{\mathrm{eq}} - f_{\mathrm{dis}}\right) \tag{5.26}$$

5.2.2.4 Dissolution of Methane Gas into Seawater

The most of methane gas is dissolved into seawater at the final stage of spill. Therefore, the estimation of total amount of dissolved methane gas takes an important role in predicting the affect of spilled hydrocarbon on the marine environment.

If the solubility of methane gas can be described by the simple Henry's law, we can calculate the dissolution mass transfer rate of methane gas as follows:

$$\frac{dm_b}{dt} = KM_g A_b\left(C_S - C\right) \tag{5.27}$$

where A_b is the surface area of bubble, C the concentration of dissolved methane gas, and C_S the saturation value of C. This formulation is valid for the condition of low pressure; however, if the pressure is high (as in deep water), the Henry's law must be modified as follows:

$$f^g = Hx^l \exp\left(\frac{pv^l}{RT}\right) \tag{5.28}$$

where f^g is the fugacity of methane gas, H the Henry's constant, x^l the mole fraction of dissolved methane gas in solution, p the partial pressure of gas, and v^l the partial molar volume of methane gas. By using this modification, the solubility C_S is also modified.

$$C_S = H^* f^g \exp\left(\frac{(1-P)v^l}{RT}\right) \tag{5.29}$$

where H^* is an another form of the Henry's law constant. The estimation of the Henry's constant is an arbitrary calculation since it depends on empirical relations (Sloan 1997; Weiss 1974; Yamamoto et al. 1976). Thus, researcher must select the most appropriate configuration according to the temperature and pressure conditions.

5.2.2.5 Lagrangian Tracking Model of Gas Bubble at the Far Field

When the control volume is at the neutral buoyancy level, the plume phase is terminated, and the tracking method must be changed to a Lagrangian-type scheme. In the plume phase, each CV has the information about gas bubbles and hydrates; thus we can change the variables of computation according to the size distribution. In the advection/diffusion phase in the far field, the governing equation of bubbles is the one that is used in the simple oil-tracking model:

$$m_p \frac{du_p}{dt} = -C_m m_f \frac{du_p}{dt} - C_D \rho_f \frac{1}{8} \pi d^2 u_p^2 - m_p g + m_f g \tag{5.30}$$

where m_p is the mass of methane gas bubble, m_f the mass of ambient water displaced by gas bubble, u_p the gas bubble velocity, C_D the drag coefficient, C_m the additional mass coefficient ($C_m = 0.5$ for rigid and spherical shape), and Re the Reynolds number based on bubble diameter. The horizontal motion of gas bubble is also represented by the random walk model, the same way as done by the oil model. The changes of mass, density, and size are calculated with the same formulation explained in the plume model.

5.3 Field Simulation

5.3.1 Deepspill Experiment

After the Deepwater Horizon (DWH) accident, numerous oil spill simulations, especially for the Gulf of Mexico, have been carried out by various researchers and companies. The DWH is a well-known example for an oil spill case, but no comprehensive in situ measurements in deep water have been done. For the modelers of oil/gas spill, one of the valuable field experiments is the "Deepspill" at the Norwegian Sea. This field experiment was conducted by the SINTEF in Norway, and the following three blowout experiments were carried out: diesel/gas (Exp-A), crude oil/gas (Exp-B), and gas/water (Exp-C) blowouts. The discharges were realized at the −844 m depth level, and the longest discharge took place for 2 h. The details of the experiment can be found in the SINTEF Report (Johansen et al. 2001); a summary of experimental conditions is given in Table 5.2. In this field experiment, the sizes of oil droplet and gas bubble are determined by a direct collection, and the concentration of spilled hydrocarbon was measured with the sound echo technique. Data on the ambient water and current flow were obtained by ADCP. Figure 5.4 shows the oil and bubble size distribution of the Deepspill

Table 5.2 Conditions of "Deepspill" experiment (Johansen et al. 2001) (Disel/gas experiment is not included here)

Parameter	Crude/gas exp.	Water/gas exp.
Label	Exp-B	Exp-C
Date	June 29	June 29
Release period	1.0 h	2.0 h
Water discharge rate	1.0 m^3/min	1.0 m^3/min
Gas discharge rate	0.7 Sm^3/s	0.7 Sm^3/s
Diameter of the orifice	0.12 m	0.12 m
Density of discharged oil	842.5 kg/m^3	–
Temperature of released mixture	−0.77 °C	−0.77 °C
Initial release velocity	2.08 m/s	2.08 m/s
Horizontal diffusion coefficient	0.01 m^2/s	0.01 m^2/s
Vertical diffusion coefficient	0.001 m^2/s	0.001 m^2/s

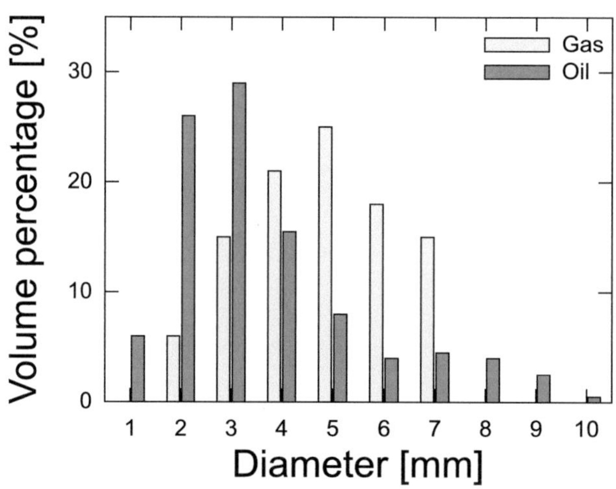

Fig. 5.4 Initial bubble size distribution of methane gas released from seabed

experiment in which the size measurements were done near the discharge points. The median values are 3 mm for oil and 5 mm for gas, and these results were well agreed with the estimation done with population balance (Bandra and Yapa 2011) and with the measurements of a laboratory experiment (Brandvik et al. 2013; Johansen et al. 2013). In the simulation, these size data are used for the initial (boundary) conditions at an inlet nozzle of discharge. Figure 5.5 shows the temperature and salinity distributions in the vertical direction in the field experiment. Although all three discharge tests were not simultaneously conducted, these profiles were not largely changed compared with the simulation time scale; thus, we used the same distribution at each time step in the present simulation. On the other hand, the temporal variation of current flow significantly affects the simulation results; thus,

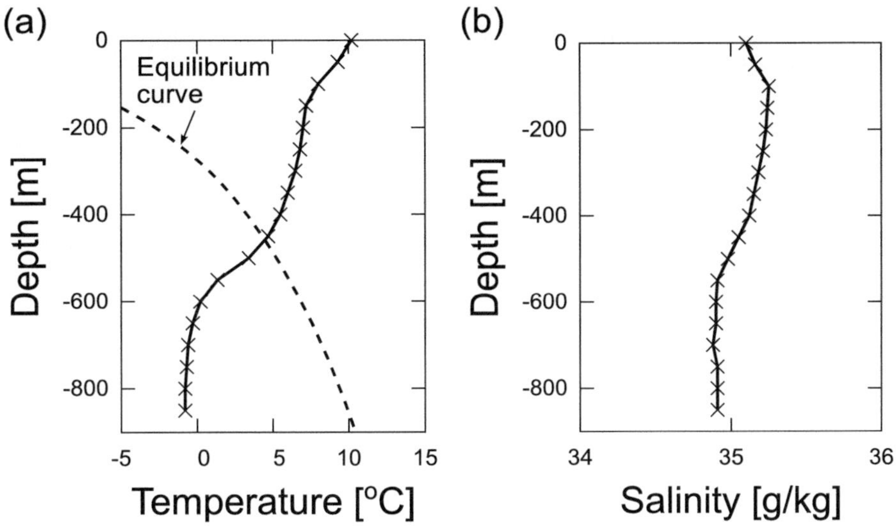

Fig. 5.5 Vertical distributions of (**a**) temperature with the equilibrium curve of methane and (**b**) salinity of seawater

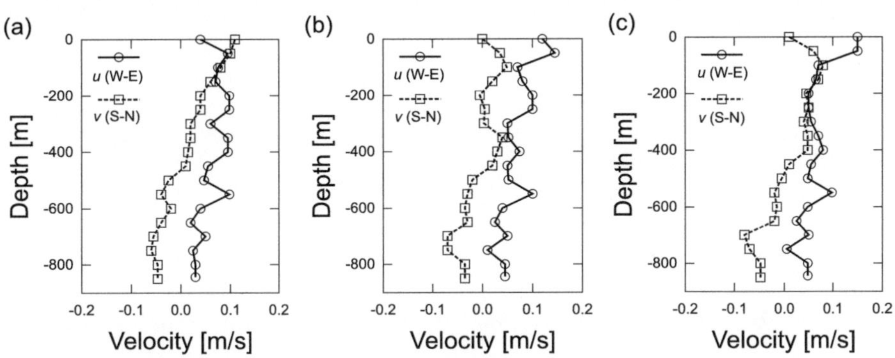

Fig. 5.6 Vertical distribution of current flow velocity for Exp-B every half hour: (**a**) $t = 0.0$ h, (**b**) $t = 0.5$ h, (**c**) $t = 1.0$ h

we had to use unsteady current flow data from the field observations or from the ocean models. The current flow velocity distributions in the Deepspill experiment are shown in Figs. 5.6 and 5.7. In the area where experiment was conducted, a strong current flow was seen at the middle depth (about −600 to −500 m below the shallow water near the sea surface), and the current flow was temporarily changed in each direction. In this chapter, we present the numerical results for the crude oil and gas discharge tests of the Deepspill experiment as seen in Table 5.2. In the simulation, the spill behavior of crude oil was determined by the simple oil-tracking model and that of gas by the hybrid plume/Lagrangian model.

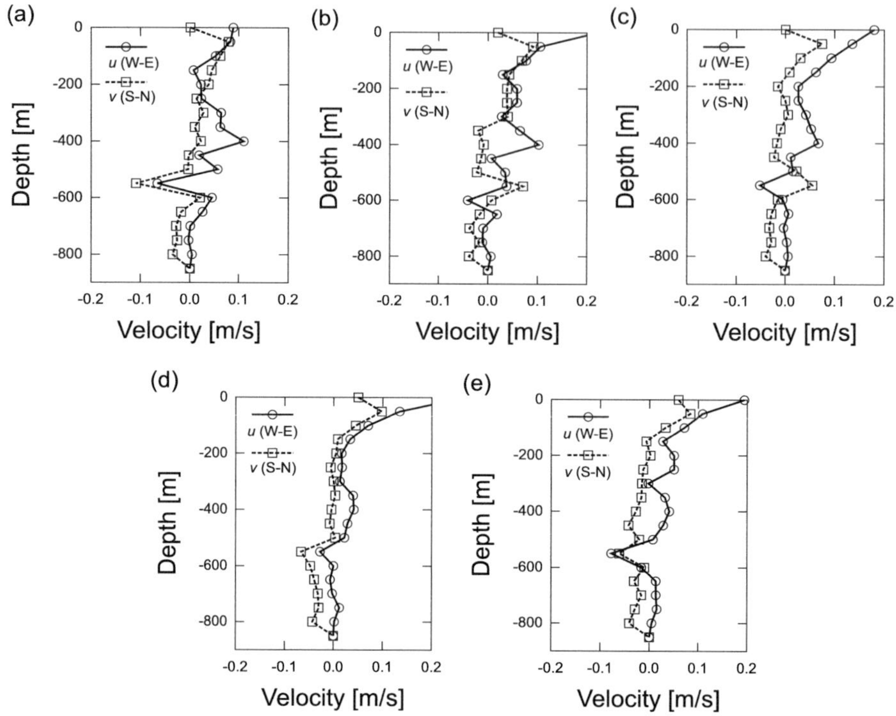

Fig. 5.7 Vertical distribution of current flow for Exp-C velocity every half hour: (**a**) $t = 0.0\,\text{h}$, (**b**) $t = 0.5\,\text{h}$, (**c**) $t = 1.0\,\text{h}$, (**d**) $t = 1.5\,\text{h}$, (**e**) $t = 2.0\,\text{h}$

5.3.2 Crude Oil Spill Behavior Without Phase Change

In Sect. 5.2, we introduced two spill models, namely, the simple oil-tracking model and the hybrid plume/Lagrangian model. In this part, the numerical simulation results for the crude oil discharge test (Exp-B) in the Deepspill model with the simple oil-tracking technique are presented. Because of our limited computational resources, the total number of released oil droplets for 1-h blowout was about 1200. This is small compared with the actual value estimated from the discharge flow rate of the experiment. Each oil particle was discharged with a constant jet velocity from the inlet nozzle, and the particle size was given randomly according to the droplet size distribution shown in Fig. 5.4. Figure 5.8 shows the outline of oil droplet trajectory in the vertical cross sections of the west-east and south-north directions after a 1-h blowout. For comparison with the experimental results, the sound eco data (Johansen et al. 2001) are also shown in Fig. 5.8. It is noted that the experiment data are accumulative for all periods, and the unsteady motion of the plume structure was not captured in the experiment. The computed distribution in the south-north direction (Fig. 5.8b) well agrees with the experimental data; however, the distribution in the west-east direction is slightly different. The difference between

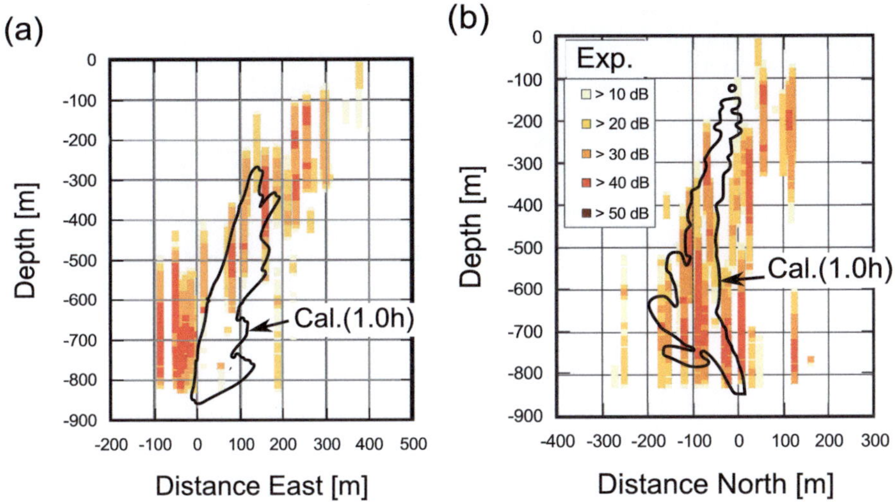

Fig. 5.8 Comparison of spilled oil distribution between the simulation with simple Lagrangian model and the field measurement: (**a**) the west-east direction and (**b**) the south-north direction

the numerical and experimental data is mainly due to the accuracy of the present numerical model. Other reasons for the difference would be the uncertainty of ambient flow conditions and measured data. As mentioned in the introduction section, many researchers have developed and validated oil-tracking models. Further details for the Lagrangian tracking model of oil droplet can be found elsewhere.

5.3.3 *Gas Spill Behavior with Phase Change*

Modeling of gas spill in deep water is continuously developing because the methane gas accompanies phase change. This makes its modeling challenging, and predictions by the conventional oil-tracking models may not be accurate.

Under the ambient flow conditions of deep water, methane exists as hydrate since the temperature and pressure conditions satisfy the requirement for methane hydration. In Fig. 5.5, the equilibrium curve of methane is shown, and it crosses the distribution at about −500 m depth. Therefore, methane gas exists as hydrate below −500 m depth thermophysically. The density of methane hydrate is larger than that of methane gas, and the rising velocity is quite different between the solid and gas phases, and the predicted behavior might also be drastically different. From this viewpoint, we carried out simulations for pure methane gas with or without methane hydration. The numerical condition is the Exp-C of the Deepspill, but the ambient current flow is not considered, that is, no current flow. Figure 5.9 shows the methane hydrate prediction for a 2-h blowout period. In the methane hydration simulation, the conventional hydrate formation model given in Eq. 5.13 was used

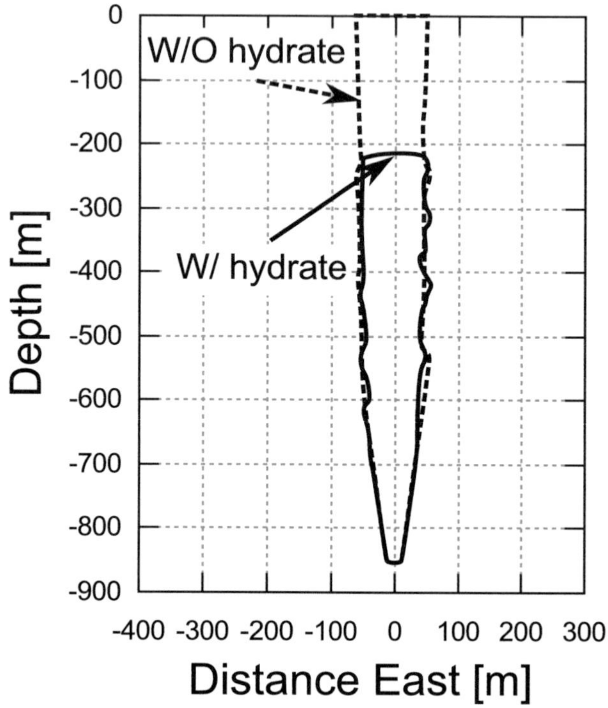

Fig. 5.9 Effect of methane hydration on the terminal plume height under no current flow

with the empirical parameter value of $\Psi = 38$. When hydration was considered, the plume was extended only up to around -200 m depth as seen from the figure. The height of this plum structure is caller terminal since it was not changed if the period of discharge was extended for more than 2 h. On the other hand, when hydration is not included in the simulation, most of the discharged methane gas reached to the sea surface due to strong buoyancy acting on gas bubbles. Results of the spilled methane show a conical plume shape below about -700 m depth, and the near/far field was automatically switched to the hybrid scheme. To investigate the behavior of methane phase change in the near field, the mass balance in the Lagrangian CVs was calculated, and the results are shown in Fig. 5.10. As the conventional hydrate formation model requires the empirical parameter Ψ to be known, we parametrically changed the value of Ψ from 1 to 150. Temporal and spatial variations of methane mass for $\Psi = 38$ and $\Psi = 150$ are presented in Fig. 5.10. As seen, the discharged methane gas gradually changed to hydrate during rising, and the total methane mass decreased due to the dissolution into the ambient seawater. Although the total methane mass predicted was not highly dependent on Ψ, the mass of generated hydrate and the consumption of methane gas were sensitive to the Ψ values. A large Ψ value indicates that the growth of methane hydrate is enhanced. Indeed, in the run using $\Psi = 150$, the methane hydrate was quickly formed, and the increase of hydrate mass was saturated within 30 s and up to the -820 m level. The range of jet/plume region at the near field was also affected by Ψ. Figure 5.11 shows the

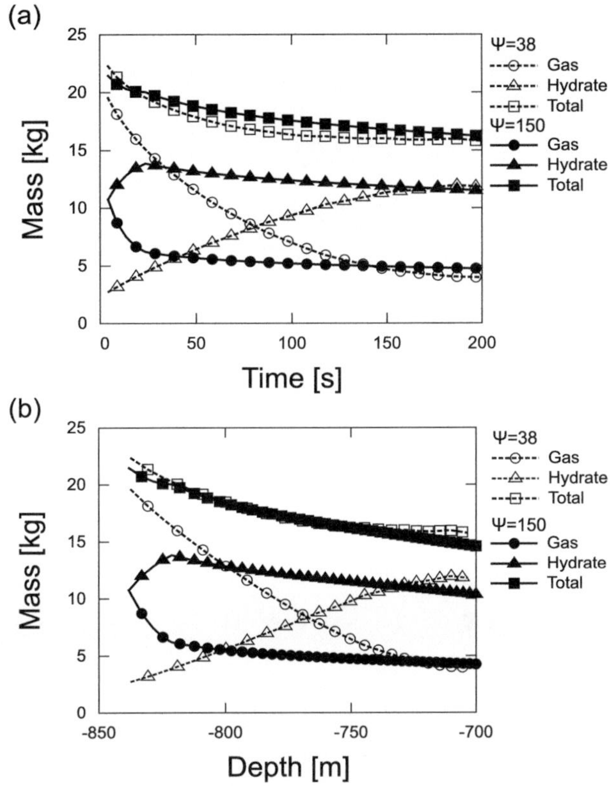

Fig. 5.10 Mass balance of methane in the plume region under no current flow: (**a**) time development, (**b**) depth dependency

dependency of Ψ on the neutral buoyancy level and the time period (arrival period) to reach to this level. Although the growth rate of hydrate is linearly proportional to the value of Ψ as seen in Eq. 5.13, the calculated NBL and the arrival period were not monotonically related, and their minimum values were around $\Psi = 38$. Chen and Yapa (2003) suggested that the proper value of Ψ is 38 in their simulation for the Exp-C of the Deepspill experiment with the CDOG model (CDOG is the same hybrid model as explained in Sect. 5.2). However, the optimum value of Ψ might change according to the field environment and blowout conditions, and the simulation without Ψ is preferable if possible.

The use of the Ranz-Marshall equation explained in Sect. 5.2 can be introduced into the hybrid model. We carried out a simulation for the Exp-C with the new model including the Ranz-Marshall equation. In order to validate the numerical results with experimental measurements, the current flow condition in Fig. 5.7 was used. Figure 5.12 shows the comparison of the calculated methane spill region with the experimental data for a 2-h blowout period. The calculated spill region was agreed with the experimental data especially in the high concentration region of plume core, and the arriving level of spilled methane was properly predicted (about −160 m). However, in the near field around the blowout point, a wide horizontal diffusion

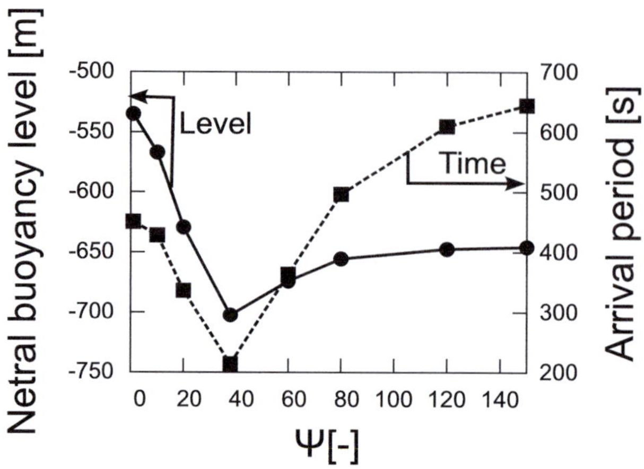

Fig. 5.11 Effect of empirical parameter (considered in the hydrate growth calculation) on the neutral buoyancy level and its arrival period

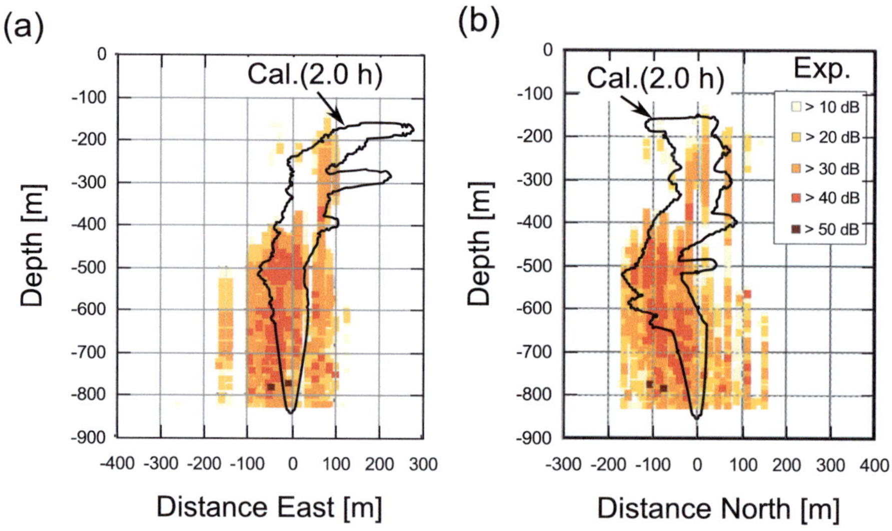

Fig. 5.12 Comparison of spilled methane distribution for the Exp-C between the simulation with the Ranz-Marshall equation and the field measurement: (**a**) the west-east direction and (**b**) the south-north direction

region of methane observed in the experiment was not simulated. Thus, some additional modeling considerations for the jet/plume region in the near field might be necessary for more accurate predictions. We carried out the same simulation with the Ranz-Marshall equation in the case of hydrate formation. Results from the models were not clearly different (figure is not shown). To discuss the effect of the Ranz-Marshall equation, the mass balance in the near field was presented in Fig. 5.13, and the result of the conventional model ($\Psi = 38$) is also shown

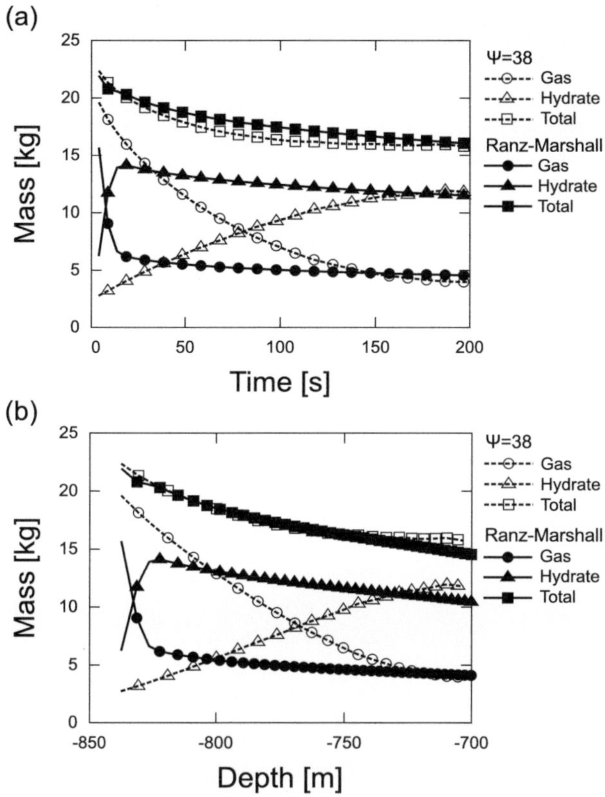

Fig. 5.13 Mass balance of methane in the plume region for the Exp-C: (**a**) time development, (**b**) depth dependency

for comparison. When the Ranz-Marshall model was used, methane hydrate was quickly formed at the initial stage of blowout the same as the result of $\Psi = 150$ under no current flow (Fig. 5.10). Thus, it is found that the enhancement of mass transfer due to the ambient convective flow was well considered in the present new model.

Finally, the simulated three-dimensional development of spilled gas for the Exp-C is presented in Fig. 5.14. At the initial stage of blowout ($t = 0.5$ h), a jet/plume structure with conical shape was formed in the near field, and the center line of the structure was not affected by the current flow because its strength was weak, almost zero at this depth level. After that, the rising methane gas/hydrate was meandering due to the effect of unsteady current flow. The formed spill structure was then branched and diffused since this phase was simulated by the Lagrangian tracking model with bubble size distribution. When the rising methane gas reached at the limit of methane gas existence at $t = 2.0$ h, the growth of spilled methane structure stopped at this stage, and the mass flow rate of injected methane was balanced with the mass of dissolved methane in the ambient water.

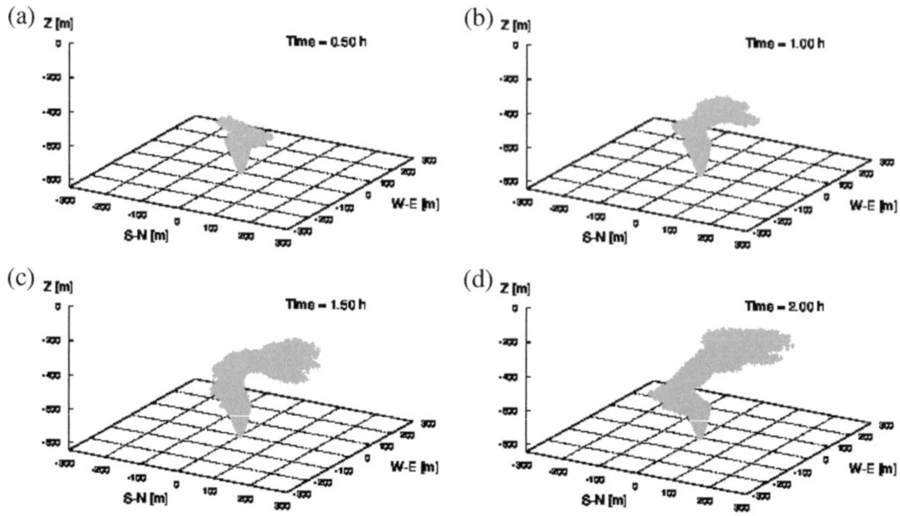

Fig. 5.14 Three-dimensional motion of formed spill structure: (**a**) $t = 0.5\,\text{h}$, (**b**) $t = 1.0\,\text{h}$, (**c**) $t = 1.5\,\text{h}$, (**d**) $t = 2.0\,\text{h}$

5.4 Summary

Two numerical models for tracking oil and gas blowout in deep water were developed based on the conventional plume and Lagrangian approaches. The simple oil-tracking model was valid if droplet breakup and coalescence were not involved. On the other hand, the tracking of methane requires various physical models and kinetics of methane hydrate, and its modeling was affected by the assumptions used for the basic gas bubble/hydrate particle. The standard hydrate shell model governed by thermophysical stability gave a reasonable result for predicting large-scale motions, but the physics of rising methane gas/hydrate was strongly affected by the empirical parameters and the small-scale modeling. In the field, oil and gas blowouts often occur together, and thus we have to consider the interaction between oil droplets and gas bubbles. Some researcher discussed and modeled the gas separation from oil plume when the plume structure is bending in the horizontal direction under strong current flow; however, the interaction between oil and gas in the plume core was not fully discussed.

A spill accident must be simulated quickly, and damages must be estimated as accurately as possible. In that respect, we think that the conventional tracking models discussed in this chapter will be powerful tools, if, particularly, they are supported by in situ measurements such as carried out by underwater robots/vehicles. More precise modeling might have been possible if we had access to significant computational resources and the required time were available.

Acknowledgements This work was financially supported by a Grant-in-Aid for Scientific Research (S) (No. 23226017) from the Ministry of Education, Culture, Sports, Science and Technology of Japan. The authors express their sincere thanks to Mr. Kawahara and Mr. Iwao for their assistance in numerical calculations.

References

Bandra UC, Yapa PD (2011) Bubble sizes, breakup, and coalescence in deepwater gas/oil plumes. J Hydraul Eng 137:729–738

Brandvik PJ et al (2013) Droplet breakup in subsurface oil releases – part 1: experimental study of droplet breakup and effectiveness of dispersant injection. Mar Pollut Bull 73:319–326

Chen F, Yapa PD (2003) A model for simulating deepwater oil and gas blowouts – part II: comparison of numerical simulation with "Deepspill" field experiments. J Hydraul Res 41:353–365

Chen F, Yapa PD (2004) Modeling gas separation from a bent deepwater oil and gas jet/plume. J Mar Syst 45:189–203

Clift R et al (1978) Bubbles, drops, and particles. Dover Publications, New York

Englezos P et al (1987) Kinetics of formation of methane and ethane gas hydrates. Chem Eng Sci 42:2647–2658

Joahnsen Ø (2000) DeepBlow – a Lagrangian plume model for deep water blowouts. Spill Sci Technol Bull 6:103–111

Johansen Ø, Rye H, Melbye AG, Jensen HV, Serigstad B, Knutsen T (2001) DeepSpill JIP – experimental discharge of gas and oil at Helland Hansen, June 2000, Technical Report. SINTEF Report STF66 F01082, SINTEF Applied Chemistry, Trondheim

Johansen Ø et al (2013) Droplet breakup in subsea oil releases – part 2: predictions of droplet size distributions with and without injection of chemical dispersants. Mar Pollut Bull 73:327–335

Matsumoto R et al (2009) Formation and collapse of gas hydrate deposits in high methane flux area of the Joetsu Basin, Eastern Margin of Japan Sea. J Geogr 118:43–71 (In Japanese with abstract in English)

McDougall TJ (1978) Bubble plumes in stratified environments. J Fluid Mech 85:655–672

North EW et al (2011) Simulating oil droplet dispersal from the deepwater horizon spill with a Lagrangian approach. Geophys Monogr Ser 195:217–226

Ranz WE, Marshall WR (1952) Evaporation from drops. Chem Eng Prog 48:141–146

Reed M et al (1999) Oil Spill Contingency And Response (OSCAR) analysis in support of environmental impact assessment offshore Namibia. Spill Sci Technol Bull 5:29–38

Seol D-G et al (2009) Measurement of behavioral properties of entrained ambient water in a stratified bubble plume. J Hydraul Eng 135:983–988

Sloan ED (1997) Clathrate hydrates of natural gases. Marcel Dekker, New York

Socolofsky SA et al (2015) Intercomparison of oil spill prediction models for accidental blowout scenarios with and without subsea chemical dispersant injection. Mar Pollut Bull 96:110–126

UNESCO (1981) Eighth report of the Joint Panel on Oceanographic Tables and Standards, Sidney, B. C., September 1980. UNESCO Technical Papers in Marine Science 36:25

Weiss RF (1974) Carbon dioxide in water and seawater: the solubility of a non-ideal gas. Mar Chem 2:203–215

Yamamoto S et al (1976) Solubility of methane in distilled water and seawater. J Chem Eng Data 21:78–80

Yapa PD et al (2008) Modeling the impact of an accidental release of methane gas in deepwater. In: Proceeding of Oceans 2008, Quebec City

Yapa PD et al (2010) A model to simulate the transport and fate of gas and hydrates released in deepwater. J Hydraul Res 48:559–572

Yapa PD et al (2012) How does oil and gas behave when released in deepwater? J Hydro-Environ Res 6:275–285

Zheng L, Yapa PD (2002) Modeling gas dissolution in deepwater oil/gas spills. J Mar Syst 31:299–309

Zheng L, Yapa PD (2003) A model for simulating deepwater oil and gas blowouts – part I: theory and model formulation. J Hydraul Res 41:339–351

Chapter 6
Effect of Liquid-Gas Interaction on Plume Structure in Blowout Flow

Tatsuya Ochi, Youhei Takagi, Yasunori Okano, and Naomi Kato

Abstract Multiphase flow simulation for oil and gas blowout in water is performed to investigate the effect of liquid-gas interaction on the plume structure. The mass and momentum balance equations were solved for each phase, and the interactions were considered with the Euler-Euler approach. To consider the interphase interaction between two dispersed phases in the oil-gas-water three-phase system, we applied the drag force between oil and gas in the momentum equation. A series of numerical results showed that the interaction between the two dispersed phases affected the velocity distribution of the gas and oil phases in the multiphase plume. These results suggested that the gas-oil interaction might affect the traveling time of gas and oil in an actual marine environment.

Keywords Hydrodynamics • Multiphase flow • CFD • Plume behavior • Interphase interaction

6.1 Introduction

To estimate the extent of oil pollution and to prevent oil spills from spreading and causing damage to marine environments, early detection and monitoring systems must be deployed around offshore oil and gas production facilities. For this purpose, spilled oil tracking autonomous buoys (SOTAB) (Kimura et al. 2013; Senga et al. 2014) and spilled oil and gas tracking methods by numerical simulation (Takagi et al. 2012; Tsutsukawa et al. 2012) have been developed. However, since field experiments and actual sea observations are limited because of environmental restrictions and economic aspects, laboratory-scale experiments have also been carried out to investigate the characteristics of oil and gas blowout flow at SINTEF (Brandvik et al. 2013; Johansen et al. 2013).

T. Ochi • Y. Takagi (✉) • Y. Okano
Department Materials Engineering Science, Osaka University, Toyonaka, Osaka, Japan
e-mail: t.ochi@cheng.es.osaka-u.ac.jp; takagi@cheng.es.osaka-u.ac.jp

N. Kato
Department Naval Architecture and Ocean Engineering, Osaka University, Suita, Osaka, Japan

© Springer Japan 2017

N. Kato (ed.), *Applications to Marine Disaster Prevention*,
DOI 10.1007/978-4-431-55991-7_6

In order to find a strategy that minimizes the damage caused by spilled oil and gas, several experimental studies and numerical models have been developed (Zheng et al. 2002; Johansen et al. 2003; Yapa et al. 2012; North et al. 2013). In addition, Fabregat et al. numerically investigated the effect of the presence of gas bubbles on the turbulent mixing and entrainment of a multiphase plume, while neglecting the gas-phase momentum (Febregat et al. 2015). Results of numerical modeling of oil and gas releases with the ADMS model (Yapa et al. 2012) showed that there could be a momentum-driven stage near the blowout site and a buoyancy-driven stage near the sea surface. In the momentum-driven stage, which is near the blowout site, not only the ambient parameters (temperature, salinity, and water current) but also the interphase interaction, such as the drag force exerted between each phase, are considered to be critical factors for the prediction of the flow behavior because of the existence of clustered oil droplets and gas bubbles.

In our previous study, which was based on the Lagrangian approach, we carried out a simulation of the overall spill from the blowout site (seabed) to the sea surface for oil and gas (Takagi et al. 2012). Our previous model adopted an integrated approach, which was composed of two methods: a Lagrangian control volume method in the near field close to the blowout site and a Lagrangian parcel tracking method in the far field.

In the Lagrangian approach, the blowout oil and gas were assumed to be a set of multiple fluid particles, and the calculated plume behavior included only the interaction between the continuous phase (water) and the dispersed phases (gas or oil). The flow fields in the previous cases were oil-water two-phase flow without a gas phase and gas-water two-phase flow without an oil phase. However, the actual flow field observed in deepwater spill accidents is oil-gas-water three-phase flow (McNutt et al. 2012). In order to improve the numerical accuracy of the previous model, the interaction between dispersed gas and oil must be considered.

The previous integrated Lagrangian model could not solve the ambient flow state and only calculated the movement of particles by the equation of motion. In contrast, computational fluid dynamics (CFD) can solve the fluid motion in the entire domain in the near field, but demands on computer memory and calculation time are large when CFD is applied for the jet/plume flow. There are two approaches for the numerical simulation of multiphase flow: the Euler-Lagrange approach and the Euler-Euler approach. In the Euler-Lagrange approach, the continuous (fluid) phase is treated as a continuum by solving the Navier-Stokes equations, while the dispersed phase is solved by tracking the particles through the calculated flow field. The dispersed phase can exchange momentum, mass, and energy with the fluid phase. However, the calculation cost becomes large as the number of particles increases. This approach is commonly used in cases where the dispersed phase occupies a low volume fraction. In the Euler-Euler approach, both the continuous phase and dispersed phase are treated as continua, and the concept of phase volume fraction is introduced since the volume of one phase cannot be occupied by the other phase. This approach is suited for high volume fractions of the dispersed phase.

In this chapter, the interaction between two dispersed phases (gas bubbles and oil droplets) was considered by utilizing the Euler-Euler approach. The volume of fluid (VOF) method and multi-fluid modeling are known for this approach. The VOF method is used when interface tracking and detailed resolution are important factors (e.g., to model the structure of the single bubble rising in water). However, it is difficult to apply to a large-scale simulation because a fine mesh is required to resolve the dispersed-phase particles. In multi-fluid modeling, each phase fraction is locally averaged in each cell, and we can resolve the characteristics of the whole plume structure with low computational cost compared to that of the VOF method. For the purpose of our study, the multi-fluid model was adopted to simulate the plume behavior.

In the present multi-fluid modeling, by regarding the fluids as continua, we calculated the transient flow state of the ambient fluid, as well as that of the plume, and investigated the interphase interactions in detail. Unlike the interaction between a continuous phase and a dispersed phase in two-phase flow, the interaction between two dispersed phases in three-phase flow has not been modeled so far because of the complexity of the flow. Several approaches for the interaction between two dispersed phases in three-phase flow have been discussed for gas-liquid-solid systems by using various assumptions (Mitra-Majumdar et al. 1997; Padial et al. 2000; Schallenberg et al. 2005), and an interaction force was added as an additional source term in the momentum balance equations of each dispersed phase. However, only a few investigations on gas-liquid-liquid systems have been carried out (e.g., oil-gas-water three-phase flow in a vertical pipe, Cazarez et al. 2010).

With this as a background, we carried out a numerical simulation for oil-gas-water three-phase blowout flow to investigate the effect of the interaction between the two dispersed phases on the plume behavior.

6.2 Numerical Method

The flow field observed in the oil and gas blowouts from the seabed is an oil-gas-water mixture, that is, the fluid, which forms the plume structure, consists of three phases. Water is considered to be the continuous phase, whereas the gas bubbles and the oil droplets are considered to be dispersed phases. To consider the interactions between each pair of phases, we applied the multi-fluid model in the Euler-Euler framework.

6.2.1 Governing Equations

The phase volume fractions satisfy the following conservative condition:

$$\alpha_w + \alpha_g + \alpha_o = 1, \tag{6.1}$$

where α is the phase fraction and the subscripts w, g, and o indicate the water, gas, and oil phase, respectively.

The continuity equations for each phase are given as:

$$\frac{\partial \alpha_k}{\partial t} + \mathbf{u}_k \cdot \nabla \alpha_k = 0, \tag{6.2}$$

where t is time and $\mathbf{u}_k$ is the velocity vector. The subscript k indicates the phase (water, gas, or oil).

The momentum equations for each phase are expressed as:

$$\begin{aligned} &\frac{\partial(\rho_k \alpha_k \mathbf{u}_k)}{\partial t} + (\rho_k \alpha_k \mathbf{u}_k \cdot \nabla)\, \mathbf{u}_k \\ &= -\alpha_k \nabla p + \nabla \cdot (\mu_k \alpha_k \nabla \mathbf{u}_k) + \rho_k \alpha_k \mathbf{g} + \mathbf{M}_{D,k}, \end{aligned} \tag{6.3}$$

where ρ is the phase density, p is the pressure, μ is the dynamic viscosity, $\mathbf{g}$ is the gravitational acceleration, and $\mathbf{M}_D$ is the momentum transfer term.

6.2.2 *Description of the Interphase Interaction*

The term is composed of the total drag force between each phase:

$$\mathbf{M}_{D,w} = -\mathbf{M}_{D,gw} - \mathbf{M}_{D,ow}, \tag{6.4}$$

$$\mathbf{M}_{D,g} = \mathbf{M}_{D,gw} - \mathbf{M}_{D,go}, \tag{6.5}$$

$$\mathbf{M}_{D,o} = \mathbf{M}_{D,ow} + \mathbf{M}_{D,go}, \tag{6.6}$$

where the subscripts gw, ow, and go denote the relations between "gas-water," "oil-water," and "gas-oil," respectively. Since the density of a gas bubble is much smaller than that of an oil droplet, we assumed that the gas phase gives momentum to the oil phase. We calculated the drag force between the continuous phase and the two dispersed phases as follows (Clift et al. 1978):

$$\mathbf{M}_{D,gw} = \frac{3}{4} \rho_w \alpha_w \alpha_g C_D \frac{|\mathbf{u}_g - \mathbf{u}_w| (\mathbf{u}_g - \mathbf{u}_w)}{d_g}, \tag{6.7}$$

$$\mathbf{M}_{D,ow} = \frac{3}{4} \rho_w \alpha_w \alpha_o C_D \frac{|\mathbf{u}_o - \mathbf{u}_w| (\mathbf{u}_o - \mathbf{u}_w)}{d_o}, \tag{6.8}$$

where C_D is the drag coefficient and d is the diameter of the gas bubble or oil droplet.

The interaction between the two dispersed phases (gas and oil phases) must be taken into account for three-phase flows, since the particles in the vicinity of bubbles tend to follow the bubbles (Mitra-Majumdar et al. 1997; Schallenberg et al. 2005).

In the jet flow region, we assumed that many particles (gas bubbles and oil droplets) are mixed in the continuous phase. Since all phases are assumed to be continua in this simulation, it is reasonable to model the interaction between oil droplets and bubbles in the same way as that between the continuous and dispersed phases. Thus, we implemented the two-dispersed interphase interaction term based on the oil- and gas-phase state. If the volume fraction of the gas is larger than that of the oil phase in a computational cell, the drag force is given as:

$$\mathbf{M}_{D,go} = \frac{3}{4}\rho_g\alpha_g\alpha_o C_D \frac{\left|\mathbf{u}_o - \mathbf{u}_g\right|\left(\mathbf{u}_o - \mathbf{u}_g\right)}{d_o}, \tag{6.10}$$

and if the gas-phase fraction is smaller than that of oil, it is given as:

$$\mathbf{M}_{D,go} = \frac{3}{4}\rho_o\alpha_o\alpha_g C_D \frac{\left|\mathbf{u}_g - \mathbf{u}_o\right|\left(\mathbf{u}_g - \mathbf{u}_o\right)}{d_g}. \tag{6.11}$$

In the following calculations, we assumed that the gas bubbles and oil droplets have spherical shapes; thus, the drag coefficient for each phase is calculated using the Schiller and Naumann model (Schiller and Naumann 1935):

$$C_D = \begin{cases} \frac{24\left(1+0.15\mathrm{Re}^{0.683}\right)}{\mathrm{Re}} & \mathrm{Re} \le 1000 \\ 0.44 & \mathrm{Re} > 1000 \end{cases}, \tag{6.12}$$

where the Reynolds number is calculated from the slip (relative) velocity between the continuous phase and the dispersed phase $|\mathrm{u}_d - \mathrm{u}_c|$, the particle diameter of the dispersed phase d_d, and the kinematic viscosity of the continuous phase ν_c:

$$\mathrm{Re} = \frac{\left|\mathbf{u}_d - \mathbf{u}_c\right| d_d}{\nu_c}. \tag{6.13}$$

When the dispersed-phase fraction of oil was smaller than that of gas in a cell, the Reynolds number was calculated as follows:

$$\mathrm{Re} = \frac{\left|\mathbf{u}_o - \mathbf{u}_g\right| d_o}{\nu_g}, \tag{6.14}$$

and when the phase fraction of oil was larger than that of the gas phase:

$$\mathrm{Re} = \frac{\left|\mathbf{u}_g - \mathbf{u}_o\right| d_g}{\nu_o}. \tag{6.15}$$

In interface-capturing methods, such as the VOF method in the Euler-Euler approach, surface tension is an important factor to calculate the interfacial curvature. In contrast, in the multi-fluid model, the interface is not captured precisely, but the phase state and movement are calculated by solving the fundamental equations. Thus, the surface tension is not considered in the calculations.

In an oil spill, the diameters of oil and gas particles strongly affect the spill behavior (Yapa et al. 2012). However, the size distributions and the breakup-coalescence effects of bubbles and droplets are neglected, and the dispersed-phase particle diameters are set to a constant diameter for each phase to focus on the extent of the drag force between each phase.

6.2.3 Numerical Procedure

The governing equations were discretized by the finite volume method, and the velocity and pressure were coupled by the pressure implicit with splitting of operator (PISO) algorithm (Issa 1986). All primitive variables were defined at the center of each control volume (cell). Values at the cell interfaces were calculated by the van Leer scheme (van Leer 1977). For the discretization of the time derivative, a first-order implicit Euler method was adopted. The numerical solutions were obtained by the open-source CFD code OpenFOAM. The solution procedure of the multiphase solver named multiphaseEulerFoam is summarized as follows (Wardle and Weller 2013):

1. Update the time step.
2. Solve the coupled set of phase volume fraction equations by Eq. 6.2.
3. Compute the drag coefficients by Eq. 6.12.
4. Construct the equation set for phase velocities by Eq. 6.3 and solve for the preliminary values.
5. Solve pressure-velocity coupling according to PISO algorithm:
   (a) Compute the mass fluxes at the discretized cell faces.
   (b) Solve the pressure equation.
   (c) Correct the fluxes.
   (d) Correct the velocities and apply the boundary conditions.
   (e) Repeat for number of PISO correction steps.
6. Compute turbulence and correct the velocities.
7. Repeat from 1 for the next time step.

Since the fluid velocity near the inlet nozzle is large, high Reynolds numbers can be locally generated. To take into account the effect of turbulence at a high phase velocity, large-eddy simulation (LES) was implemented in the multiphase solver, and the standard Smagorinsky model was used for the sub-grid scale (Smagorinsky 1963).

6.3 Results and Discussion

6.3.1 *Validation of the Numerical Code: Three-Phase Bubble Column*

There are only a few benchmark problems for multiphase flow including three phase, especially for gas-liquid-liquid systems. To focus on the interaction between the two dispersed phases, we used a three-dimensional bubble column reactor including gas, liquid, and solid phases (Cui and Fan 2004; Jia et al. 2007) as a validation of the present numerical code. The schematic diagram of the experimental setup is shown in Fig. 6.1, and the dimensions and the computational grid number are summarized in Table 6.1. The physical properties of each phase are shown in Table 6.2. Air, water, and alginate gel beads were used as the gas, liquid, and solid phase, respectively. The value of the solid-phase viscosity is set to the same as that of water, since its variations will not significantly affect the simulation results

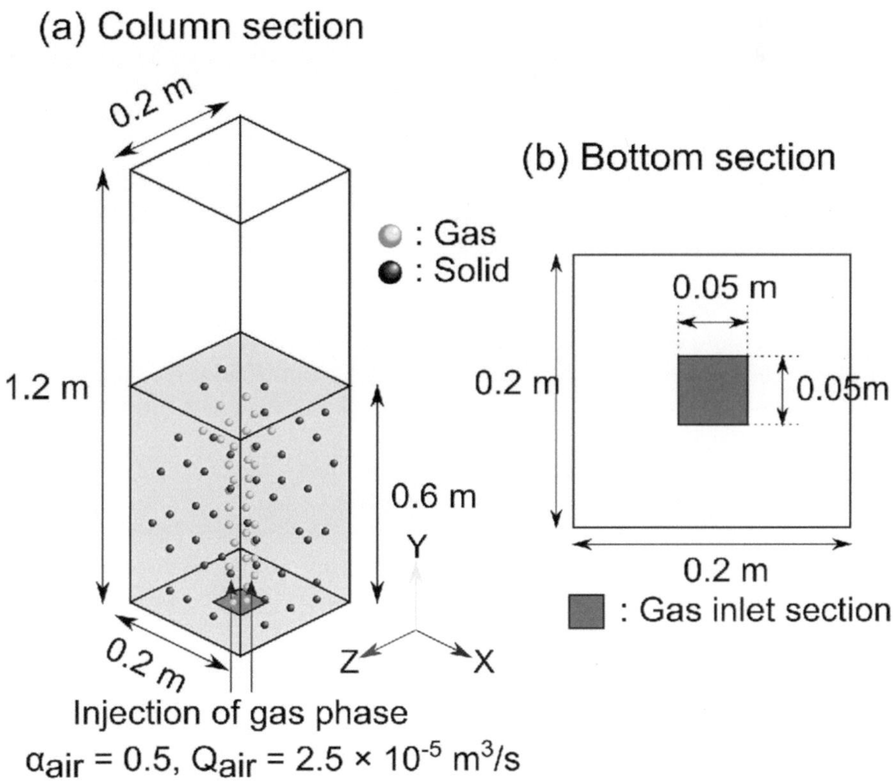

Fig. 6.1 Schematic diagram of gas-liquid-solid three-phase bubble column reactor (Jia et al. 2007): (**a**) column section, (**b**) bottom section and inlet configuration

Table 6.1 Dimensions and computational grid numbers of the three-phase bubble column: X is the width, Y is the height, Z is the depth, and N_x, N_y, and N_z are the computational grid numbers in each direction

Properties	Values
Column dimension	
$X \times Y \times Z$	$0.2 \times 1.2 \times 0.2$ m
Grid resolution	
$N_x \times N_y \times N_z$	$40 \times 80 \times 40$
$\Delta x \times \Delta y \times \Delta z$	$5 \times 15 \times 5$ mm
Inlet section	
$X \times Z$	0.05×0.05 m

Table 6.2 Physical properties of air, water, and solid phase

Phase	Density	Kinetic viscosity	Diameter
Air	1.18 kg/m^3	1.84×10^{-5} m^2/s	4.5 mm
Water	997 kg/m^3	0.89×10^{-6} m^2/s	–
Solid	1048 kg/m^3	–	3.5 mm

(Schallenberg et al. 2005). The bubble column was initially filled with stationary tap water to a height of $h = 0.6$ m from the bottom. Transient calculations were started assuming that the solid phase was distributed uniformly in the reactor and the solid loading of α_{solid} was 2.0 vol.%. Air bubbles were injected at a constant flow rate of 2.5×10^{-5} m^3/s, and the gas volume fraction at the inlet section was 0.5. The Reynolds number based on the inlet length and the superficial velocity of gas bubbles at the inlet section was 27. The bubble diameter required in the drag force calculation was considered constant with a value of 4.5 mm, and the diameter of the solid particles was 3.5 mm, corresponding to the experiment. The boundary conditions on the walls were no slip for the liquid (water) phase and free slip for the gas (air) and solid phases. The top boundary was an opening, and only the air phase could leave the numerical domain.

Figure 6.2 shows the phase distribution of gas and oil at $t = 60$ s after the gas injection. The gas phase formed a vertical plume structure and released out of the numerical domain from the top boundary. Solid particles were well mixed in the bubble column because of the flows of the gas and water phases.

After the gas bubbles reached the liquid surface, the plume structure periodically meandered in the horizontal direction. This is demonstrated by Fig. 6.3, which shows the time trend of the lateral water velocity at the center point of $h = 0.5$ m ($X = 0.1$ m, $Y = 0.5$ m, $Z = 0.1$ m). After approximately 40–50 s elapsed, the plume began meandering horizontally. As the velocity took positive and negative values, the plume swung to the right and left sides, respectively. According to spectrum analyses for the horizontal water velocities, the meandering period with a long wavelength was 6.8 s for the case with interaction of the two dispersed phases and 14.7 s for the case without interaction. The motions with these periods correspond to the oscillation period of the plume structures.

Figure 6.4 shows the comparison of the lateral distribution of the time-averaged axial water velocities calculated by the present numerical code with experimental data at the column height of $h = 0.5$ m in the central plane ($X = 0$–0.2 m, $Y = 0.5$ m,

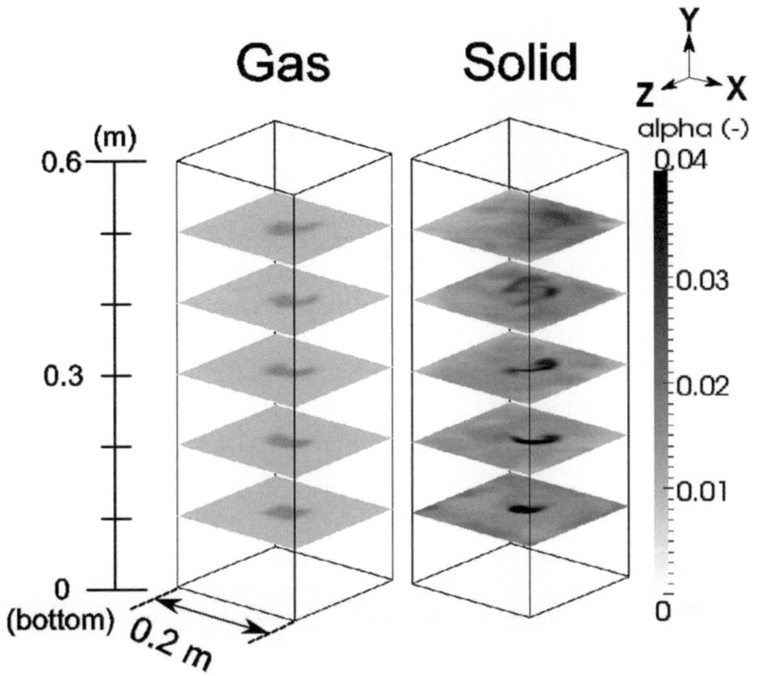

Fig. 6.2 Phase distribution of the gas- and solid-phase volume fractions of the column at each column height (0.1–0.5 m from the bottom section) in the case with dispersed phase-dispersed phase interaction $\mathbf{M}_{D,go}$ at time $t = 60$ s after the gas injection

$Z = 0.1$ m). The axial water velocities were temporally averaged over 120 s after reaching the fully developed state. Both calculation cases were in qualitative agreement with the experimental results, and the case with the interaction term between the two dispersed phases, $\mathbf{M}_{D,go}$, showed better agreement at the center of the column and in the vicinity of the walls than the case without $\mathbf{M}_{D,go}$. The characteristic of the plume structure was well simulated by the present numerical code, and the implementation of the interaction between the two dispersed phases was efficient for predicting the three-phase flow behavior. The present simulation was also used in the following gas and oil blowout flow.

6.3.2 Effect of the Interaction Between Gas and Oil Phases on the Plume Structure: Oil-Gas-Water Three-Phase Flow Behavior

It is difficult to obtain a steady-state flow for the blowout of a liquid-gas mixture in water since the accumulation of injected components strongly affects the flow

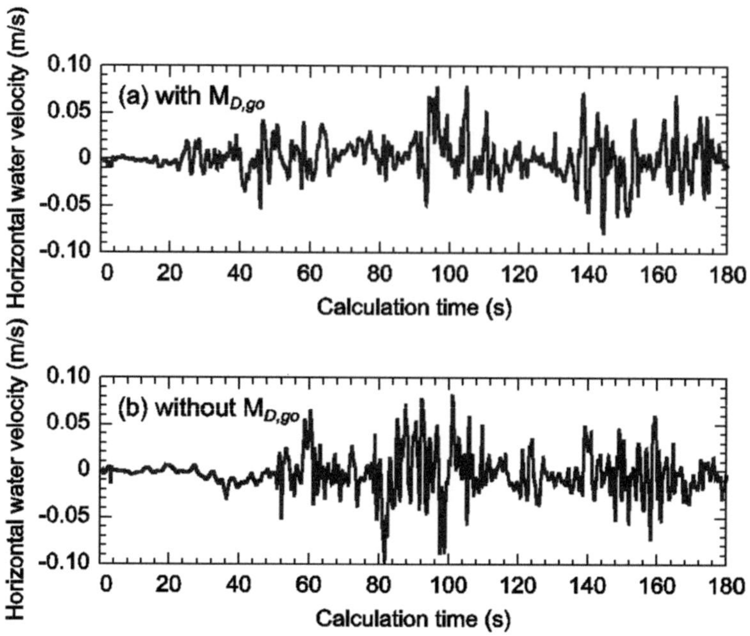

Fig. 6.3 Time development of the horizontal water velocity at the central point of $h = 0.5$ m: (**a**) with two-dispersed-phase interaction, $\mathbf{M}_{D,go}$, (**b**) without one

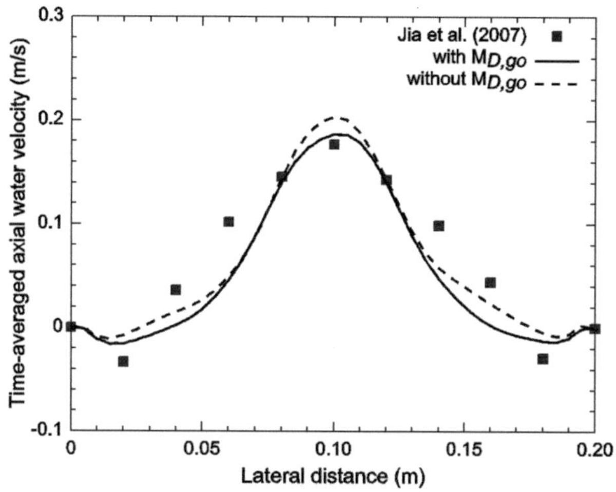

Fig. 6.4 Time-averaged axial water velocity at $h = 0.5$ m in the central plane: *red square* point represents the experimental data (Jia et al. 2007); *dashed line* represents the calculation results of the present numerical code without the drag force between gas bubbles and oil solid particle, $\mathbf{M}_{D,go}$; and the *solid line* represents the case with $\mathbf{M}_{D,go}$

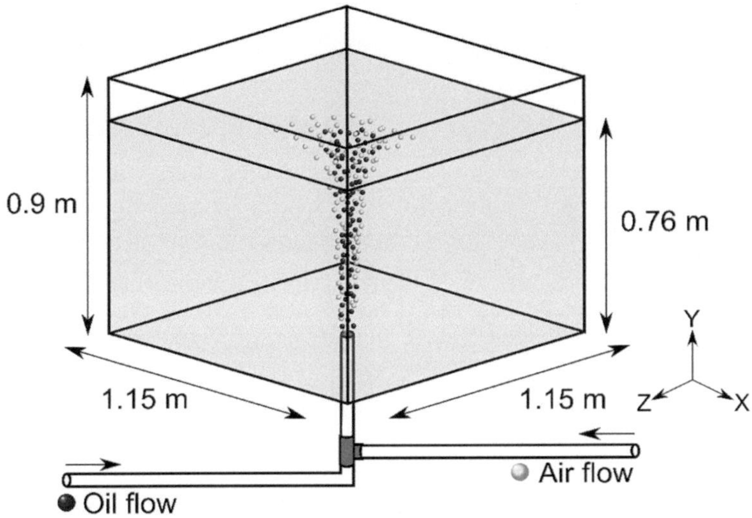

Fig. 6.5 Schematic diagram of the square water tank (Neto et al. 2008)

structure in a small vessel. To investigate the effect of the interaction between gas and oil phases on the plume structure and behavior, we carried out a gas and oil blowout flow simulation in a 1.15 × 1.15 m square water tank, which is similar to the one used in a previous gas and water jet flow experiment (Lima Neto et al. 2008).

Figure 6.5 shows a schematic diagram of the square water tank for the present jet flow calculation. The computational mesh is nonuniform and numerical cells were clustered near the jet axis. Although there was a tendency for the numerical results to converge on each primitive variable, the validity of the grid resolution could not be confirmed because of the limited computational power. The present work used the LES turbulence model, which resolves large eddies directly, while eddies smaller than the grid scale are modeled through a filtering process. The smaller the eddy size, the smaller the effect of the eddy on the flow field, so increasing the number of grid divisions generally leads to a numerical solution with a higher resolution. In the multi-fluid model, however, grid refinement does not necessarily provide good convergence. The viewpoint of this chapter is the effect of the additional interactions between the two dispersed phases in the three-phase flow rather than the grid dependency for numerical simulation of jet flow. Thus, we investigated the three-phase flow behavior by using the following mesh. The dimensions of the tank, the inlet nozzle, and the number of cells are shown in Table 6.3. The dimension of the smallest cell, which was located along the jet axis, was $(\delta x, \delta y, \delta z) = (2$ mm, 15 mm, 2 mm). The maximum grid size was $(\delta x, \delta y, \delta z) = (15$ mm, 15 mm, 15 mm), which occurred along the side wall. The physical properties of air and water were the same as those used in the three-phase bubble column calculation in the previous validation. The oil is Oseberg from the Norwegian Sea, and its properties are shown in Table 6.4. The gas bubbles and oil droplets were assumed to have a constant

Table 6.3 Dimensions and computational grid numbers of the square water tank

Properties	Values
Water tank dimension	
$X \times Y \times Z$	$1.15 \times 0.9 \times 1.15$ m
Grid resolution	
$N_x \times N_y \times N_z$	$74 \times 60 \times 74$
$\Delta x \times \Delta y \times \Delta z$ (min)	$2 \times 15 \times 2$ mm
$\Delta x \times \Delta y \times \Delta z$ (max)	$15 \times 15 \times 15$ mm
Inlet section	
$X \times Z$	0.05×0.05 m

Table 6.4 Physical properties of oil phase (Oseberg)

Properties	Values
Oil	
Density	852.2 kg/m^3
Kinetic viscosity	9.0×10^{-3} m^2/s

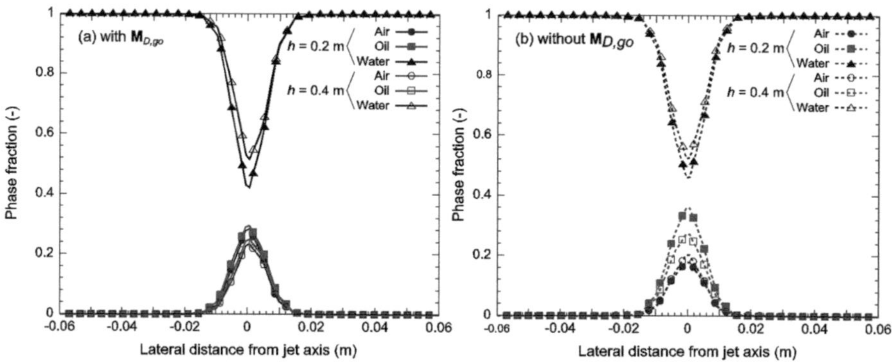

Fig. 6.6 Distribution of the instantaneous phase fraction in the lateral direction at $h = 0.4$ m: (**a**) the case with gas-oil interaction, (**b**) the case without gas-oil interaction

diameter of 3 mm. Both gas bubbles and oil droplets were injected simultaneously from the inlet section at a constant flow rate of 5.0×10^{-5} m^3/s. The phase fractions of gas and oil phase at the inlet were set to 0.5. Unlike the air-water-solid three-phase flow in the previous section, the gas- and oil-phase boundary conditions at the walls were defined as no slip.

Figure 6.6 shows the phase distribution in the central region of the water tank at time $t = 40$ s after the oil and gas injection. The lateral distance corresponds to the distance from the center of the tank. The air- and oil-phase fractions took their maximum values at the center of the plume, and water was also present in the plume core region because of the entrainment effect of the plume in both cases. In the case without gas-oil interaction, more clustered oil phase rose in the plume center region without being dragged by the high slip velocity of the gas phase.

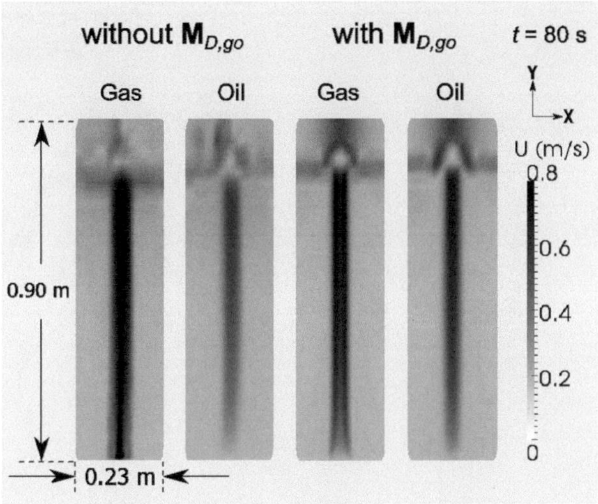

Fig. 6.7 Vertical distribution of the instantaneous velocity magnitude for gas and oil phase at the central plane ($Z = 0.575$ m): $t = 80$ s after gas and oil injection into the water tank

Figure 6.7 shows the velocity distribution of each phase at $t = 80$ s after the mixture injection. Each figure shows the center region of the tank as a part of the whole numerical domain. The gas-phase velocity was larger than that of the oil phase because of the different densities of the two phases. As a result, a slip (relative) velocity between the gas and oil phases was observed. When the effect of the interaction of dispersed gas and dispersed oil was taken into account, the velocity distribution of the gas phase did not take its maximum value at the plume core region. This is because oil phase clustered at the plume core region and exerted a drag on the gas phase, because of which the gas rising velocity was locally decreased.

Figure 6.8 shows the velocities of the gas and oil phases at $h = 0.4$ m from the bottom. The phase velocity was averaged over 40 s after reaching a fully developed state. In the case with gas-oil interaction, the gas-phase velocity was reduced, and the oil buoyant velocity increased compared to the case where the interaction was neglected. Unlike the velocity profile of the simple bubbly jet flow shown in Fig. 6.4, the time-averaged maximum gas velocity did not peak at the center of the plume because of the simultaneous gas- and oil-phase injection.

We also investigated the effect of the oil and gas particle diameters on the plume structure in the case with interaction between oil and gas. Figure 6.9 shows the vertical velocity distribution of each phase at the tank height of $h = 0.4$ m in the center plane. The initial and boundary conditions were similar to the above oil-gas jet flow except for the dimensions of the gas bubbles and oil droplets. We calculated two cases: oil droplet diameter of 0.5 mm and gas bubble diameter of 5.0 mm (case 1) and oil droplet diameter of 5.0 mm and gas bubble diameter of 0.5 mm (case 2). In case 1, the plume was meandering in the right direction because of the effect of the large gas bubbles at $t = 40$ s. Moreover, the maximum velocity of oil was smaller than the respective value in case 2 because of the increase in the

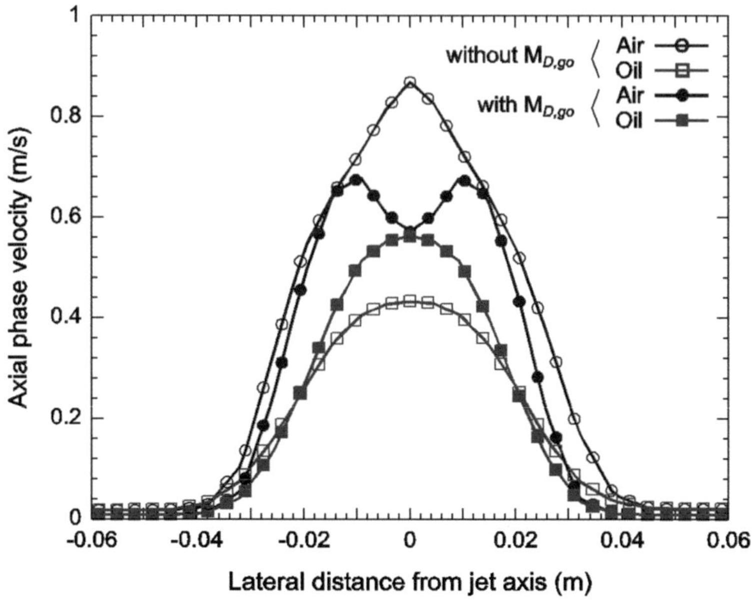

Fig. 6.8 Distributions of the time-averaged phase rising velocities at $h = 0.4$ m: *dashed lines* represent the case without the interaction between gas and oil phase $\mathbf{M}_{D,go}$, and *solid lines* represent the case with $\mathbf{M}_{D,go}$

gas-oil interaction. In addition, the velocity profile of each phase was spread in the horizontal direction compared to case 2. This is because by decreasing the gas-oil interaction because of the smaller oil droplets, the effect of the high viscosity of oil decreased. This result suggested that the upward velocity of the plume, which consists of small oil droplets and large gas bubbles, became small, and the time for the oil to drift in the sea was longer than in the case of the plume consisting of large oil droplets and small bubbles.

6.4 Conclusions

To investigate the effect of the interaction between liquid and gas in multiphase flow, especially between dispersed gas bubbles and dispersed oil droplets in three-phase flow, on the plume structure and behavior, we carried out a numerical simulation of three-phase jet flow. The interaction between the two dispersed phases (gas and oil) was additionally implemented as a drag force, and the characteristics of the multiphase plume were simulated qualitatively. The numerical results showed that the interaction between gas and oil affected the rising velocity of each phase. Oil droplets were dragged by the gas bubbles, which rose at a high velocity, and as a result, the oil buoyant velocity was larger than in the case without gas-oil interaction.

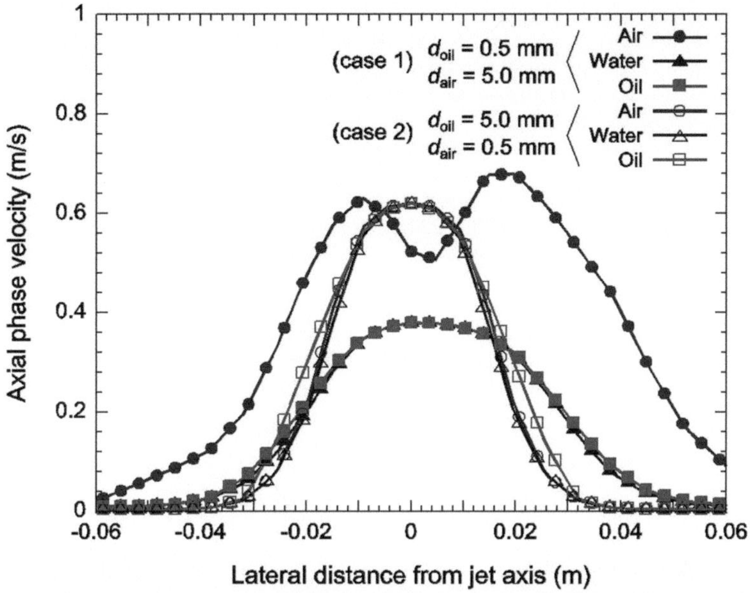

Fig. 6.9 Axial velocity distribution of each phase at $h = 0.4$ m ($t = 40$ s) including the interaction between oil and gas: (case 1) oil droplet diameter (d_{oil}) = 0.5 mm and gas bubble diameter (d_{air}) = 5.0 mm and (case 2) d_{oil} = 5.0 mm and d_{air} = 0.5 mm

The diameters of the oil and gas particles were found to also affect the spill behavior, especially in terms of the velocity profiles of each phase.

In the actual field, the gas phase mainly consists of methane, and mass transport (e.g., gas dissolution) and kinetic reaction processes (e.g., hydrate reaction) may affect the plume structure at deepwater conditions of low temperature and high pressure. The previous numerical model considered the effects of mass transport and some kinetics on the gas phase. Therefore, it is an important task to investigate the effect of gas-oil interaction on those processes by implementing the interaction between the gas and oil phases, to ultimately improve the numerical accuracy of the conventional Lagrangian model.

Acknowledgments This research project is being funded for 2011FY–2015FY by Grant-in-Aid for Scientific Research(S) of Japan Society for the Promotion of Science (No. 23226017).

References

Brandvik PJ et al (2013) Droplet breakup in subsurface oil releases – part 1: experimental study of droplet breakup and effectiveness of dispersant injection. Mar Pollut Bull 73:319–326

Cazarez O et al (2010) Modeling of three-phase heavy oil-water-gas bubbly flow in upward vertical pipes. Int J Multiphase Flow 36:439–448

Clift R et al (1978) Bubbles, drops and particles. Academic, New York
Cui Z, Fan LS (2004) Turbulence energy distributions in bubbling gas-liquid and gas-liquid-solid flow systems. Chem Eng Sci 59:1755–1766
Febregat A et al (2015) Numerical simulations of turbulent thermal, bubble and hybrid plumes. Ocean Model 90:16–28
Issa RI (1986) Solution of implicitly discretized fluid flow equations by operator-splitting. J Comput Phys 62:40–65
Jia X, Wen J et al (2007) Local hydrodynamics modeling of a gas-liquid-solid three-phase bubble column. Am Inst Chem Eng 53:2221–2231
Johansen Ø et al (2003) DeepSpill – field study of a simulated oil and gas blowout in deep water. Spill Sci Technol Bull 8:433–443
Johansen Ø et al (2013) Droplet breakup in subsea oil releases – part 2: predictions of drop size distributions with and without injection of chemical dispersants. Mar Pollut Bull 73:327–335
Kimura R et al (2013) Guidance and control of an autonomous underwater robot for tracking and monitoring spilled plumes of oil and gas from seabed. In: Proceedings of international offshore and polar engineering conference. pp 366–371
Lima Neto IE et al (2008) Bubbly jets in stagnant water. Int J Multiphase Flow 34:1130–1141
McNutt M,K et al (2012) Review of flow estimates of the deepwater horizon oil spill. Proc Natl Acad Sci U S A 109:20260–20267
Mitra-Majumdar D et al (1997) Hydrodynamic modeling of three-phase flows through a vertical column. Chem Eng Sci 52:4485–4497
North EW et al (2013) Simulating oil droplet dispersal from the deepwater horizon spill with a lagrangian approach. Geophys Monogr Ser 195:217–226
Oil Properties Database, Environmental Science and Technology Centre, Canada. http://www.etc-cte.ec.gc.ca/databases/Oilproperties/
Padial NT et al (2000) Three-dimensional simulation of a three-phase draft-tube bubble column. Chem Eng Sci 55:3261–3273
Schallenberg J et al (2005) The important role of local dispersed phase hold-ups for the calculation of three-phase bubble columns. Chem Eng Sci 60:6027–6033
Schiller L, Naumann Z (1935) A drag coefficient correlation. Z Ver Dtsch Ing 77:318–320
Senga H et al (2014) Field experiments and new design of a spilled oil tracking autonomous buoy. J Mar Sci Technol 19:90–102
Smagorinsky J (1963) General circulation experiments with the primitive equations, monthly. Weather Rev 91:99–164
Takagi Y et al (2012) Numerical tracking of methane gas/hydrate and oil droplet in deep water spill. In: Proceedings of Inter-Academia 2012
Tsutsukawa S et al (2012) Efficacy evaluation of data assimilation for simulation method of spilled oil drifting. In: Proceedings of 5th PAAMES and AMEC2012, SEPAS-05
van Leer B (1977) Towards the ultimate conservative difference scheme IV: a new approach to numerical convection. J Comput Phys 23:276–299
Wardle KE, Weller HG (2013) Hybrid multiphase CFD solver for coupled dispersed/segregated flows in liquid-liquid extraction. Int J Chem Eng 2013:128936
Yapa PD et al (2012) How does oil and gas behave when released in deepwater? J Hydro Environ Res 6:275–285
Zheng L et al (2002) A model for simulating deepwater oil and gas blowouts – part I: theory and model formulation. J Hydraul Res 41:339–351

Chapter 7
Numerical Simulation of Spilled Oil Drifting with Data Assimilation

Hiroyoshi Suzuki

Abstract Recently, oil spill accidents from tankers and oil production facilities have been a fairly regular occurrence. When spilled oil drifts to the coast, it destroys the natural environment and ecosystem and inflicts damage on the regional environment. In order to minimize such damages, it is necessary to retrieve the oil from the sea and to place oil control equipment on the coast where the spilled oil is predicted to drift ashore.

As a result, researchers have begun to develop forecasting models of drifting oil. In order to forecast the drifting behavior of spilled oil numerically, ocean and atmospheric models are important. In addition, an oil drifting model that can describe the changing features of the oil should be described. In this research, the Princeton Ocean Model (POM) and Regional Ocean Model System (ROMS) are used as ocean models and the Weather Research and Forecasting (WRF) Model as the atmospheric model to develop an oil drifting model. Meanwhile, the data assimilation (DA) scheme is widely used in the field of weather forecasting, and this scheme is effective in improving the prediction of weather.

Given this, if the DA scheme can be applied to the spilled oil drifting simulation, the accuracy of oil behavior predictions would be improved. The application of the DA scheme to the oil drifting simulation and evaluation of its efficacy are described in this chapter.

Keywords Oil drifting simulation • WRF • POM • ROMS • Data assimilation • Oil drifting model

7.1 Introduction

Oil spill accidents caused by marine tanker mishaps and problems with offshore oil production facilities, while on the decrease, continue to occur around the world and in Japan. Examples include the Amoco Cadiz oil spill (1978: France), Exxon

H. Suzuki (✉)
Department of Naval architecture and ocean engineering, Osaka University, 2-1, Yamada-oka, Suita 565-0871, Osaka, Japan
e-mail: suzuki_h@naoe.eng.osaka-u.ac.jp

© Springer Japan 2017

N. Kato (ed.), *Applications to Marine Disaster Prevention*,
DOI 10.1007/978-4-431-55991-7_7

Valdez oil spill (1989: Alaska, USA), Nakhodka oil spill (1997: Sea of Japan), Hebei Spirit oil spill (2007: Yellow Sea, Korea), Deepwater Horizon oil spill (2010: Gulf of Mexico, USA), and so on.

When spilled oil drifts to the coast, it destroys the natural environment and ecosystem and inflicts heavy damages on the regional environment. In the case of Japan, the Nakhodka oil spill accident occurred off the coast of the Shimane Prefecture in 1997, severely damaging a wide range of Japanese coast.

In order to minimize such damage, efforts are made to retrieve the spilled oil in the sea and to place oil control equipment on the coast where oil drift is expected. To help address the problem, we have begun the development of a new spilled-oil-and-gas tracking autonomous buoy system, as well as an application for a marine disaster-prevention system named spilled oil tracking autonomous buoy system (SOTAB). The system consists of two kinds: buoy robots and an oil spill simulation.

In measure 1, the underwater buoy robot called SOTAB-I autonomously tracks and monitors the plumes of oil and gas leaking from the seabed. In measure 2, the multiple floating buoy robots called SOTAB-II autonomously track the spilled oil on the sea surface and transfer useful data to a land station through satellites in real time. In measure 3, the simulation forecasts the oil drift, incorporating the real-time measured data obtained from the SOTAB-I and SOTAB-II robots, introduced in previous chapters.

This chapter deals with a part of measure 3. Here, we introduce the development of the system to forecast the drifting behavior of spilled oil. The Princeton Ocean Model (POM) and Regional Ocean Model System (ROMS) are used for the ocean model, and the Weather Research and Forecasting (WRF) Model is used for the atmospheric model. In order to improve the accuracy of the oil drift prediction, a data assimilation (DA) scheme is also applied. The observations for the DA consider real observations and data from the SOTAB-IIs. In reality, SOTAB-II is still under development. However, it is important to evaluate the efficacy of DA in the oil drifting simulation before completing the SOTAB-II. In this simulation, the Nakhodka oil spill, which occurred in 1997 in the Sea of Japan, is used as a test case. We selected this event because data for rough observations of the behavior of spilled oil on the sea were available, together with measurements of the amount of oil that drifted ashore on the coasts of the Japanese prefectures. Based on these observed data, we evaluate the accuracy and/or efficacy of the simulation model with DA and report the results.

7.2 Nakhodka Oil Spill Accident

In this research, we use the Nakhodka oil spill as the test case of the simulation as previously mentioned. We outline the accident briefly.

Predawn on January 2, 1997, in the Sea of Japan about 106 km NNE off the Oki Islands of Shimane Prefecture, the Russian tanker Nakhodka (20000dwt; age of vessel, 26 years) with a cargo of about 19,000 kl of heavy oil, broke up in stormy weather (wind velocity about 20 m/s and wave height about 6 m). Although the captain died, the 31-person crew took refuge in life rafts and was saved. In this accident, 6679 kiloliters (kl) of oil were spilled. Moreover, 1984 kl of oil were spilled as the bow continued to drift on the sea. The trajectory of the bow of the Nakhodka is shown in Fig. 7.1

Newspapers and agencies observed and recorded the behavior and distribution of the spilled oil over time. Figures were drawn based on this information and published in a book. Figure 7.2 includes some examples. In these figures, the hatched regions on the map represent the drifting oil. It should be noted that the captions of Fig. 7.2 were originally in Japanese and were translated into English by the authors; captions have been added to aid reader with understanding.

From January 7 to 10, the spilled oil was widely distributed off the coast of Hyogo, Kyoto, Fukui, and Ishikawa prefectures. During this period, no oil was observed east of the Noto Peninsula. On January 8, some of the spilled oil had already drifted ashore on the coast of Fukui and Ishikawa prefectures. From January 11 to 13, in addition to the originally affected prefectures, spilled oil was distributed off the coast of Tottori prefecture. At the north marine area of Noto Peninsula, the oil progressed day by day to the east along the Noto Peninsula. By January 14–17, spilled oil had drifted ashore widely at the Fukui and Ishikawa prefectures. Some had crossed the Noto Peninsula and approached off the coast of Niigata Prefecture. Spilled oil drifted ashore at the Hyogo, Kyoto, Fukui, and Ishikawa prefectures on January 18–20. Almost all of the drifting oil was observed off the coast of Kyoto, Fukui, and Niigata prefectures, although some also appeared near the north part of Ishikawa Prefecture. As time passed, from January 21 to 24, oil drifted ashore at Tottori, Hyogo, Kyoto, Fukui, Ishikawa, and Niigata prefectures. Almost all of the

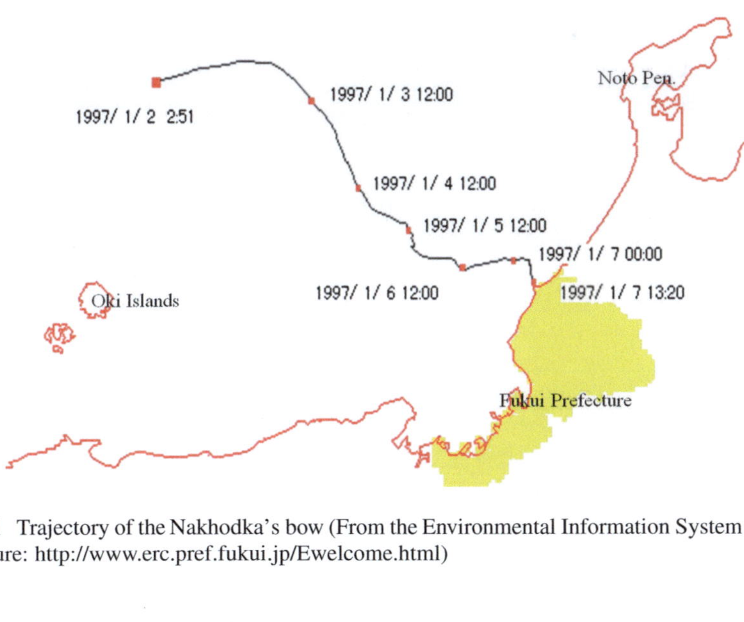

Fig. 7.1 Trajectory of the Nakhodka's bow (From the Environmental Information System of Fukui Prefecture: http://www.erc.pref.fukui.jp/Ewelcome.html)

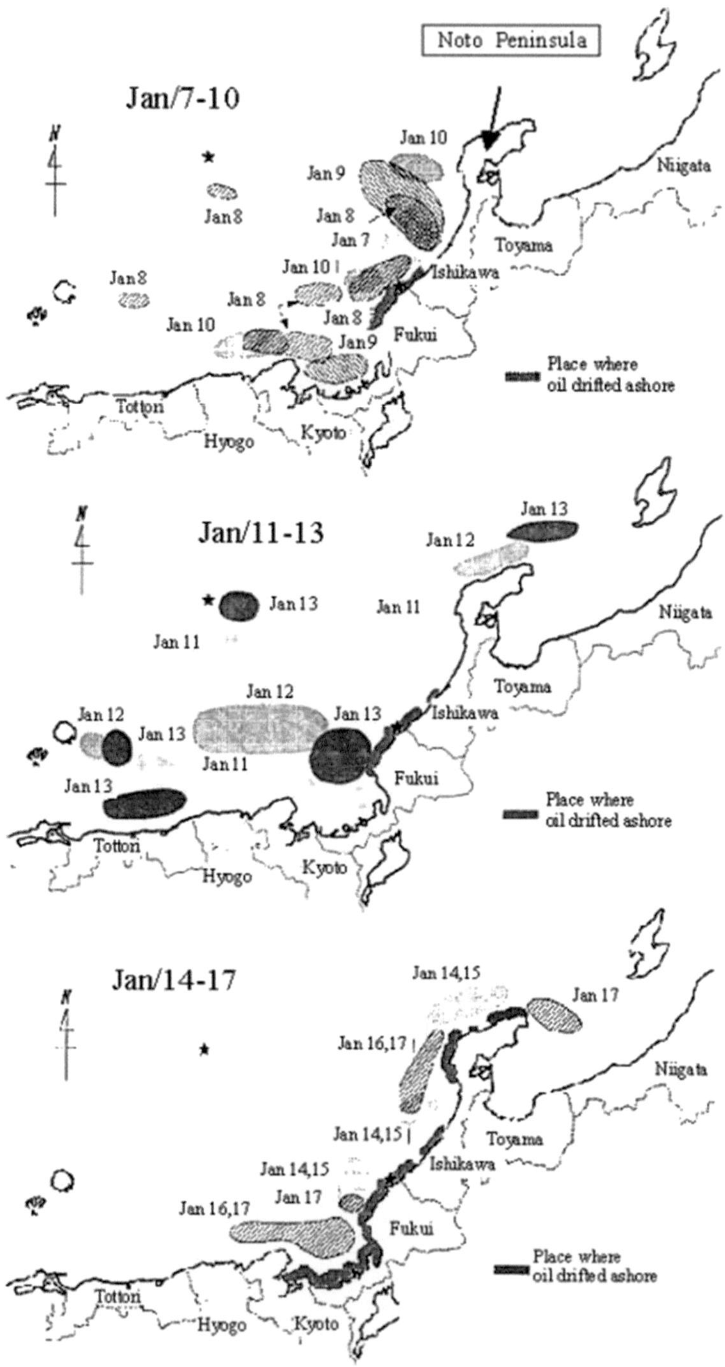

Fig. 7.2 Observed distribution of oil spilled by the Nakhodka accident with time progression (Ishii et al. 1998)

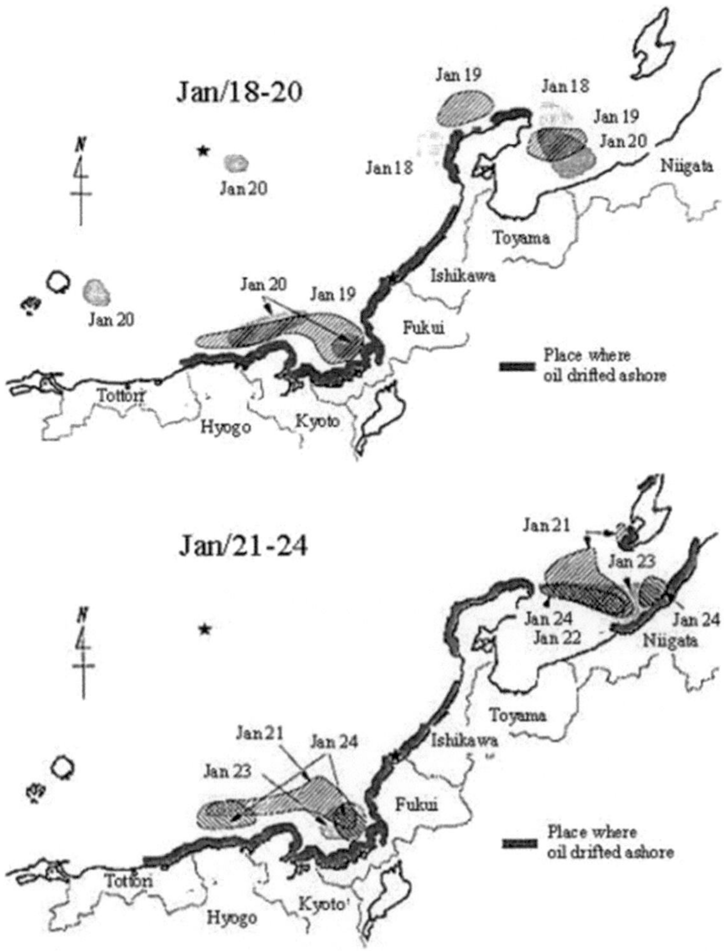

Fig. 7.2 (continued)

oil on the sea was distributed off the coast of Hyogo, Kyoto, and Fukui prefectures, as well as off of Niigata Prefecture.

Eventually, all the spilled oil from the Nakhodka was recovered from the Japanese coast facing the Sea of Japan. The amount of oil recovered from each prefecture is listed in Table 7.1. About 80 % of the spilled oil drifted ashore at the coast of the Ishikawa and Fukui prefectures. The total amount of the oil shown in Table 7.1 is different from the amount described before; this is because spilled oil absorbs water and thus increases in volume by about 8.4 times. For reference, the name of prefectures and related locations are shown in Fig. 7.3.

Table 7.1 Amount of spilled oil that drifted ashore and was recovered, by prefecture, in the Nakhodka oil spill accident (Ishii et al. 1998)

Prefectures	Amount (kl)	Percentage
Akita	225	0.5 %
Yamagata	1	0.0 %
Niigata	3795	7.6 %
Toyama	2	0.0 %
Ishikawa	22,152	44.3 %
Fukui	18,608	37.3 %
Kyoto	3614	7.2 %
Hyogo	1425	2.9 %
Tottori	70	0.1 %
Shimane	8	0.0 %

Fig. 7.3 Affected prefectures and related locations in the Nakhodka accident

7.3 Simulation of Oil Drift

7.3.1 Velocity Configuration of Spilled Oil

The velocity of drifting oil on the sea surface, U_i, can be expressed as follows:

$$U_i = Uw_i + Ut_i + Uo_i + Ud_i + Us_i \; (i = 1, 2), \tag{7.1}$$

where Uw_i, Ut_i, and Uo_i are the current velocity due to wind, the tidal current velocity on the sea surface, and the current velocity on the sea surface, respectively. Ud_i and Us_i are the velocity of oil diffusion and of oil spreading, respectively. Uw_i is basically derived from the atmospheric model WRF computation, and $Ut_i$and Uo_i are computed by POM or ROMS. The spilled oil drifting model is applied to estimate Ud_i and Us_i.

7.3.2 Treatment of Oil

Oil spilled on the sea will absorb water. Its volume will peak in the first 12 h, and its evaporative rate will also rise quickly within 24 h, as it becomes an oil mousse. Once an oil mousse is formed, it is difficult to change because of its high viscosity. Accordingly, we treated oil mousses as lumps of oil in the simulation.

In the simulation, we substituted a circular oil spot for the oil for 24 h and calculated the evaporative rate, emulsion, and spreading width. The change of heavy oil was calculated using experimental results from the Petroleum Association of Japan (http://www.pcs.gr.jp/doc/keijihenka/edatabase.html). The water content ratio was calculated according to the relationship proposed by Mackay and Zagorski (1982). Finally, the oil spread was calculated based on Fay's theory (Fay et al. 1971).

After 24 h, the particles of oil were arranged equally in a circle of that width. The turbulent diffusion caused by the oil itself was expressed using the random walk technique. The details of the random walk technique are given as

$$Ud_i = \begin{pmatrix} R_{n1} \\ R_{n2} \end{pmatrix} \sqrt{\frac{2D_H}{\Delta t}} \tag{7.2}$$

where $Rn_1$and Rn_2 are independent normalized random numbers, D_H is the horizontal diffusion coefficient, and Δt represents the time step. D_H is calculated by the Smagorinsky model (Smagrinsky 1963) as follows:

$$D_H = c^2 \Delta x \Delta y \left\{ \left(\frac{\partial u}{\partial y} \right)^2 + \left(\frac{\partial v}{\partial x} \right)^2 + \frac{1}{2} \left(\frac{\partial u}{\partial y} + \frac{\partial v}{\partial x} \right)^2 \right\}^{\frac{1}{2}} \tag{7.3}$$

where c is the Smagorinsky diffusion coefficient (=0.2), Δx and Δy are the mesh width, and u and v are the flow velocity on the sea surface.

The particles were then affected by wind and sea current. The calculation flow of the oil drifting model, including the advection effect, is shown in Fig. 7.4.

In the case of the Nakhodka accident, 6,679 kl of oil were spilled. An additional 1,984 kl were spilled while the bow drifted on the sea. In this study, we set the trajectory of the bow to the simulation as shown in Fig. 7.1 (http://www.erc.pref.fukui.jp/news/oil.html). According to the bow's trajectory, the 1,984 kl of oil were evenly dispersed. Specifically, 1 kl of oil was treated as a particle in this simulation. Therefore, 6,679 particles were positioned where the accident occurred, and 45 particles were positioned every 3 h, according to the bow's trajectory. The total number of oil spot particles was thus 8,663.

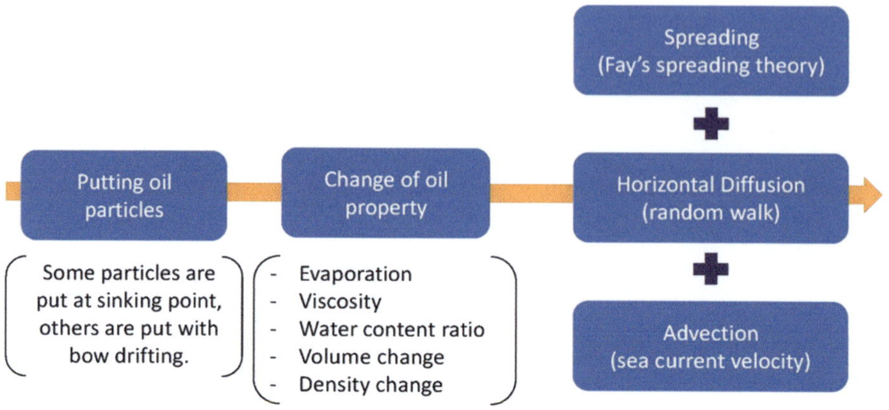

Fig. 7.4 Calculation flow of oil drifting model

7.3.3 Numerical Models of Ocean and Atmospheres

As mentioned above, we used POM and ROMS for the ocean model and WRF as the atmospheric model to evaluate the accuracy and efficacy of the numerical model using data from the Nakhodka oil spill.

The POM is one of the major ocean models. It was originally developed at Princeton University in collaboration with Dynalysis of Princeton in the early 1980s. It has been widely used all over the world. The ROMS is another important ocean model developed by the Institute of Marine and Coastal Sciences at Rutgers University. It is a free-surface, terrain-following, primitive equation ocean model widely used by the scientific community for a diverse range of applications. Meanwhile, the WRF is a numerical weather prediction (NWP) system designed to serve both atmospheric research and operational forecasting needs. The effort to develop the WRF began in the latter part of the 1990s and was a collaborative partnership principally between the National Center for Atmospheric Research (NCAR), the National Oceanic and Atmospheric Administration (represented by the National Centers for Environmental Prediction [NCEP] and the [then] Forecast Systems Laboratory [FSL]), the Air Force Weather Agency (AFWA), the Naval Research Laboratory (NRL), the University of Oklahoma (OU), and the Federal Aviation Administration (FAA).

There are two methods for estimating the advection velocity of spilled oil by the wind effect. One is the wind coefficient method, and the other is the bulk method. The wind coefficient method estimates the advection velocity directly from the wind velocity. On the other hand, the bulk method estimates wind shear stress on the sea surface from the computational results of atmospheric models and computes the flow velocity of the seawater at the surface using wind shear stress treated as the boundary condition. Uw_i can be computed from the results of the computation of the ocean model. The applied bulk method is shown below:

$$\begin{aligned} H &= c\rho C_H (U - U_s)(T_s - T) \\ E &= \rho C_E (U - U_s)(Q_s - Q) \\ \tau &= \rho C_D (U - U_s)^2 \end{aligned} \tag{7.4}$$

where τ, H, and E are stress, heat flux, and moisture flux on the sea surface, respectively; c and ρ are the isobaric specific heat of air and density of the air, respectively; U, T, and Q indicate wind velocity, temperature, and specific humidity (suffix s means "surface"), respectively; and C_D, C_H, and C_E are the transfer coefficients. The bulk method is generally used in WRF (Fairall et al. 1996).

Although a previous study (Matsuzaki and Fujita 2012) recommended using the wind coefficient method to estimate the advection velocity of spilled oil, the bulk method was applied to express wind effect here. This was because we plan to apply a DA scheme to the ocean model in future research using the data from SOTAB-IIs. SOTAB-II autonomously tracks spilled oil by controlling its direction and velocity. This means that the motion of SOTAB-II and spilled oil will be almost the same. The current velocity observed on the sea surface by SOTAB-II includes wind shear effect. To apply DA for the ocean model, the velocity field consists of Uw_i, Ut_i, and Uo_i. Therefore, we elected to use the bulk method.

7.3.4 Selection of DA Scheme

Before applying DA to the spilled oil drifting simulation, numerical experiments were carried out to check the efficacy of some computational options and to select a DA scheme. In this simulation, the DA scheme was applied only for the atmospherics model. This was because the Nakhodka oil spill occurred in the Sea of Japan during the winter, when it is well known that there are very strong winds. Therefore, we checked the efficacy of the WRF with WRFDA.

In WRFDA, several kinds of DA schemes can be implemented, including three-dimensional variational (3D-Var) method, four-dimensional variational (4D-Var) method, and ensemble Kalman Filter (EnKF). We selected 3D-Var and 4D-Var for this evaluation.

In 3D-Var, the initial value of the computation is generated by minimizing the cost function $J(x)$ and only using the observations of the time near the initial value. Equation 7.5 gives the cost function for 3D-Var:

$$J(\boldsymbol{x}) = \frac{1}{2}\left(\boldsymbol{x} - \boldsymbol{x}^b\right)^T \boldsymbol{B}^{-1}\left(\boldsymbol{x} - \boldsymbol{x}^b\right) + \frac{1}{2}(\boldsymbol{H}(\boldsymbol{x}) - \boldsymbol{y})^T \boldsymbol{R}^{-1}(\boldsymbol{H}(\boldsymbol{x}) - \boldsymbol{y}) \cdot \tag{7.5}$$

where x, x^{b}, B, y, R, and $H(x)$ are the analysis value, previous forecast, background error covariance, observation, observation error covariance, and observation operator, respectively.

On the other hand, in 4D-Var, the initial value for the computation is generated by minimizing the cost function shown in Eq. 7.6 using observations that are continued timewise:

$$J(\boldsymbol{x}) = \frac{1}{2}\left(\boldsymbol{x}_0 - \boldsymbol{x}^b\right)^T \boldsymbol{B}^{-1}\left(\boldsymbol{x}_0 - \boldsymbol{x}^b\right) + \frac{1}{2}\sum_{t=0}^{T}\left\{\boldsymbol{H}_t\left[\boldsymbol{M}_t(\boldsymbol{x}_0)\right] - \boldsymbol{y}_t\right\}^T \boldsymbol{R}^{-1}\left\{\boldsymbol{H}_t\left[\boldsymbol{M}_t(\boldsymbol{x}_0)\right] - \boldsymbol{y}_t\right\} \quad (7.6)$$

where $\boldsymbol{M}$ is the predefined model forecast operator from the initial value and the others are basically the same as in the 3D-Var DA. Generally, the 4D-Var DA is more accurate than the 3D-Var DA.

In the 3D-Var and 4D-Var DA, the definition/calculation method of the background error covariance $\boldsymbol{B}$ shown in Eqs. 7.5 and 7.6 is also selectable. In this computation, we selected and checked the cv3 and cv5 options in WRFDA. In the case of cv3, the background error covariance matrix $\boldsymbol{B}$ is calculated by the following equation:

$$\boldsymbol{B} = \left(\boldsymbol{V}\boldsymbol{B}_Z\boldsymbol{B}_X\boldsymbol{B}_Y\right)\left(\boldsymbol{V}\boldsymbol{B}_Z\boldsymbol{B}_X\boldsymbol{B}_Y\right)^T \quad (7.7)$$

where $\boldsymbol{V}$ is the standard deviation of the model, and $\boldsymbol{B}_X$, $\boldsymbol{B}_Y$, and $\boldsymbol{B}_Z$ are two horizontal components and a vertical component of the recursive filter, respectively. Then cv3 can apply the arbitrary domain of the computation.

In the case of cv5, $\boldsymbol{B}$ is defined as follows:

$$\mathbf{B} = \overline{\boldsymbol{\varepsilon}\boldsymbol{\varepsilon}^T} \cong \overline{\boldsymbol{x}'\boldsymbol{x}'^T} \quad (7.8)$$

where $\boldsymbol{\varepsilon}$ represents the true background error, although this cannot be known in reality, and x' is a model-state perturbation. It is known that x' is assumed to be statistically well represented by ε.

The NMC method, which calculates x' from the difference of two forecasts, was applied to calculate x. $\boldsymbol{B}$ by cv5 depends on the computational region. Therefore, $\boldsymbol{B}$ should be calculated using past statistics. In addition, the NMC method has been adopted in most of the weather forecast agencies around the world. A more detailed description of these option parameters may be found on the WRF website (http://www.mmm.ucar.edu/wrf/users/wrfda/).

The computational domains and computational conditions for these numerical experiments are shown in Fig. 7.5 and Table 7.2.

As the observations for the DA, we used NCEP ADP Global Surface Observational Weather Data. Specifically, METAR, SYNOP, and SHIP data were used for the DA. The observation points are shown in Fig. 7.6.

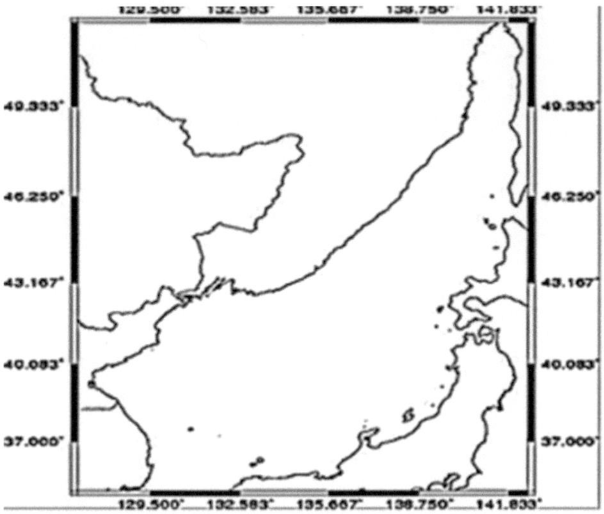

Fig. 7.5 Computational domain for numerical experiment to select DA scheme

Table 7.2 Computational conditions for numerical experiments

Model		WRF	WRFDA
Calculation domain	North latitude	34°∼ 53°	34°∼ 53°
	East longitude	127°∼142.5°	127°∼142.5°
Grid size		18km×18km	18km×18km
Time step		Δt=60(sec)	Δt=120(sec)(4DVAR)
Assimilation time window			+30 mins(3DVAR)
			+6 hours (4DVAR)
Calculation period		11/16 ~11/17,2011	11/16 ~11/17,2011
Initial value and Boundary value		NCEP Reanalysis: 2.5 degree resolution, every 6 hours	
Observed value		NCEP ADP Global Surface Observational Weather Data	

Here, METAR, SYNOP, and SHIP refer to the weather reporting formula; they show routine aviation weather reports, surface synoptic observations (the meteorological observation result from unmanned and manned observatories), and ship surface observation reports, respectively. The observed value of SYNOP, for example, is shown in Table 7.3.

Applying these methods and data, we carried out the numerical experiments. The computational results of wind velocity at 140° east longitude are shown in Fig. 7.7.

Here, the observation means the WRF solution applies detailed observation data of the NCEP Final Analysis: 1 degree of resolution, every 6 h.

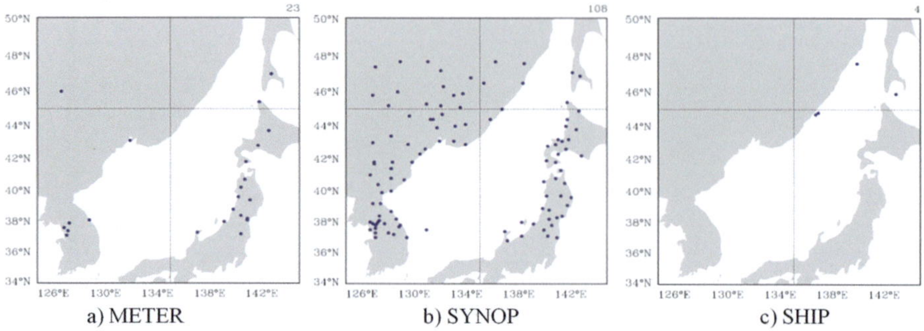

Fig. 7.6 Observation points of METER, SYNOP, and SHIP on November 16, 2011

Table 7.3 Observed data of SYNOP

Air temperature	Cloud amount/frequency	Cloud base	Cloud types
Dew point temperature	Precipitation amount	Sea level pressure	Sea surface temperature
Snow	Station height	Surface pressure	Surface winds
Swells	Visibility	Wave height	Wave period

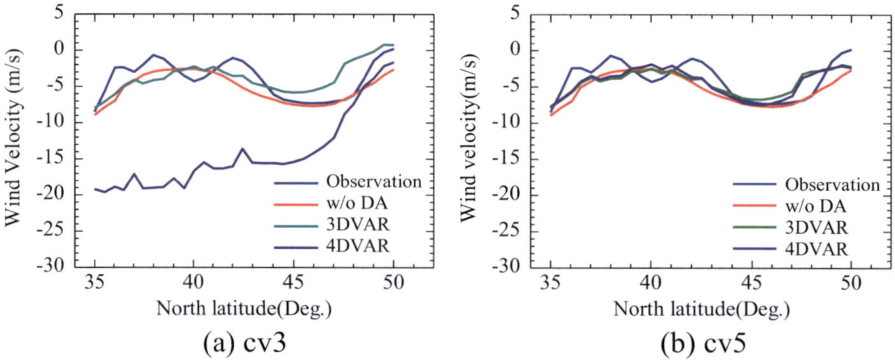

Fig. 7.7 Computational results of (**a**) cv3 and (**b**) cv5 on 140° East longitude

Based on this, the computational velocity distributions of observations, without DA, 3D-Var (cv3), 3D-Var (cv5), and 4D-Var (cv3), were similar to observed ones, except in the case of 4D-Var (cv3). Moreover, the differences between computations were small.

Next, the root mean square error (RMSE) was calculated, defined as the following expression:

$$\mathrm{RMSE} = \sqrt{\sum_{1}^{N} \frac{(x_o - x_c)^2}{N}} \tag{7.9}$$

Table 7.4 Computational results for DA test evaluated by RMSE along 140° east longitude

cv3	Without DA	3D-Var	4D-Var
RMSE	1.98	2.24	11.71
cv5	Without DA	3D-Var	4D-Var
RMSE	1.98	1.86	1.76

Table 7.5 RMSE of all grid points

	Latitude direction	Longitude direction
Without DA	2.82	2.41
3D-Var (cv3)	2.77	2.50
3D-Var (cv5)	2.73	2.26
4D-Var (cv3)	5.15	5.12
4D-Var (cv5)	2.71	2.21

Table 7.6 Improved rates of computation

	Improved rates (%)
3D-Var (cv3)	41.40
3D-Var (cv5)	43.76
4D-Var (cv3)	3.98
4D-Var (cv5)	48.20

where $x0$ is the observed values, x_c is the calculated values, and N is the number of data. Using this, the computational results of wind velocity at 140° east longitude were reevaluated. The results are shown in Table 7.4.

From these results, we find that, except for the result of 4D-Var (cv3), computational accuracy was improved by application of the DA scheme.

Table 7.5 shows the RMSE of 24-h time variation of the wind at a height of 10 m compared on all grid points. In addition, improved rates (grid points at which computational accuracy improved all grid points) based on RMSE defined as the 24-h time variation of the wind at a 10-m height were calculated and compared. The results are shown in Table 7.6

3D-Var (cv3) and (cv5) and 4D-Var (cv5) can improve computational accuracy. These results demonstrated that the DA scheme definitely improved the accuracy of the computation; in particular, 3D-Var (cv5) and 4D-Var (cv5) showed good agreement with observations.

From the viewpoint of computational load, the required number of iterations to identify the optimum value by the descent method was counted. The results are shown in Table 7.7. Here, when the results of cv3 were compared to cv5, cv5 had less iteration. In particular, the computational load of 4D-Var (cv5) was much less than that of 4D-Var (cv3).

These results indicated that the computational results of 3D-Var are similar to 4D-Var when applying parameter cv5. In other words, we considered the computational accuracy of 3D-Var (cv3), 3D-Var (cv5), and 4D-Var (cv5) to be almost the same. On the other hand, the computational load of cv5 was less than

Table 7.7 Iterations of calculation

	Iterations
3D-Var (cv3)	14
3D-Var (cv5)	5
4D-Var (cv3)	50
4D-Var (cv5)	8

that of cv3. Therefore, from the viewpoint of computational load and numerical accuracy, we chose 3D-Var (cv5) as the DA scheme.

7.4 Efficacy of DA for Estimation of Oil Drift

7.4.1 Prediction of Amount of Oil Drifted Ashore at a Certain Time Using POM and WRF (Tsustukawa et al. 2012)

For the real operation of this system, one of the primary objectives is to predict the behavior of drifting oil on the sea surface, especially the time, place, and quantity of the spilled oil that will drift ashore at a certain point in time after the accident. Moreover, the prediction should be calculated as soon as possible after the event.

On the other hand, because oil drifting phenomena proceed according to time, the position of the oil, and meteorological phenomena, the oceanographic phenomena change every moment. From this viewpoint, simulations to predict when, where, and how much spilled oil will drift ashore when the accident occurs and at certain points in time thereafter are carried out to compare the validity of 3D-Var and 4D-Var. In the case of the Nakhodka oil spill, the accident occurred on January 2, 1997, and almost all of the heavy oil had drifted ashore to the coasts of the prefectures facing the Sea of Japan by January 31, 1997. Therefore, the five starting time points for the simulation to predict the oil drifting ashore are as follows: (1) the time of the accident (January 2), (2) 5 days after the accident (January 7), (3) 12 days after, (4) 18 days after, and (5) 21 days after. A conceptual diagram of this simulation is shown in Fig. 7.8.

In this simulation, POM and WRF were used for ocean model and atmospheric model, respectively. Computational domain and condition are shown in Fig. 7.9 and Table 7.8. For the computation, we used 18-km mesh for the ocean and atmospheric models.

Using these numerical models and conditions, two case simulations were carried out without a DA scheme but applying 3D-Var (cv5) DA. As a reference, one case was computed applying 4D-Var (cv5) as a DA scheme. In the cases without DA computation, initial value of oceanographic and atmospheric data was set at each starting time, and drifting oil distribution as of January 31 was predicted. In the case with DA computation, oceanographic and atmospheric data was set at each starting time, and the DA scheme was applied every 6 h after the computation began to

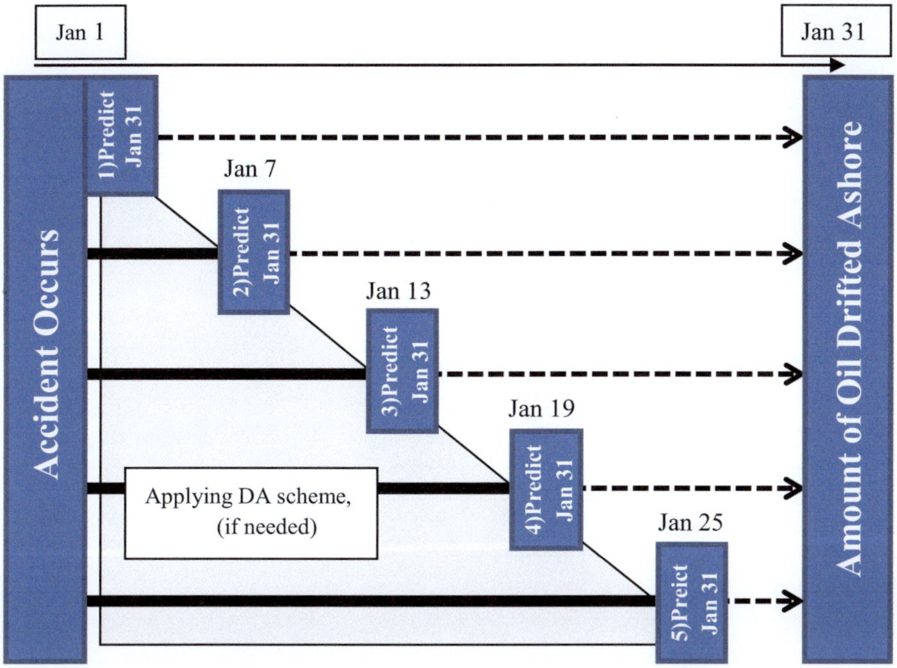

Fig. 7.8 Conceptual diagram of the simulation

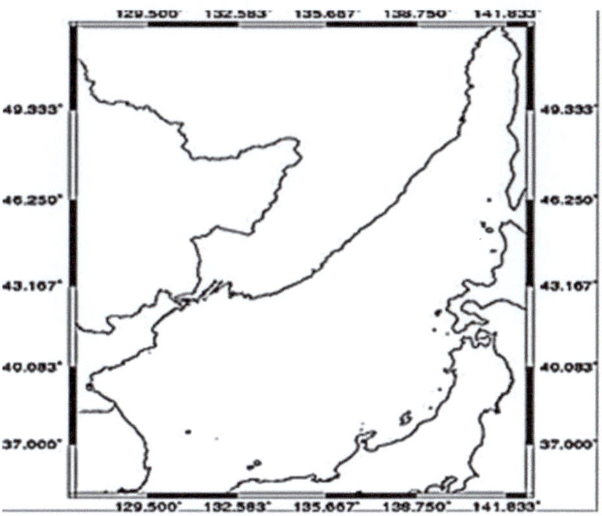

Fig. 7.9 Computational domain for POM and WRF

Table 7.8 Computational conditions for the simulation with POM and WRF

Model		WRF	WRFDA	POM
Calculation domain	North latitude	34°~ 53°	34°~ 53°	34°~ 53°
	East longitudinal	127°~142.5°	127°~142.5°	127°~142.5°
Grid size		18km×18km	18km×18km	18km×18km
Time step		Δt=60(sec)	Δt=120(sec) (4D-VAR)	Δt=12(sec)
Assimilation time window			+30 minutes (3D-VAR)	
			+6 hours (4D-VAR)	
Calculation period		1/1 ,1997~ 1/31,1997	1/1 ,1997~ 1/31,1997	1/1 ,1997~ 1/31,1997
Initial value and Boundary value		NCEP Reanalysis: 2.5 degree resolution, every 6 hours		Tide level Water temperature Salinity Forced flow
Observed value		· NCEP ADP Operational Global Surface Observations · NCEP ADP Operational Global Upper Air Observations		

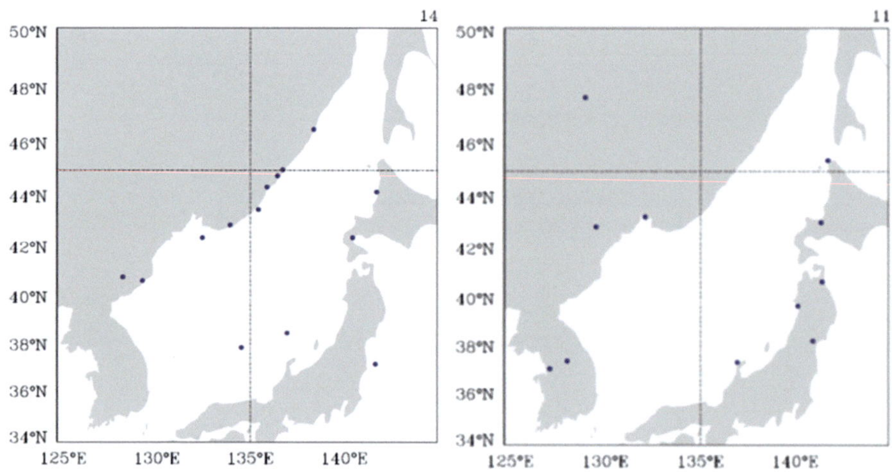

a) SYNOP b) SOUND

Fig. 7.10 Observation points of SYNOP and SOUND at 0:00, January 2, 1997

predict drifting oil behavior on January 31. The SYNOP and SOUND observation data were used for the DA computation. Here, SOUND refers to observations by radiosondes.

As the Nakhodka oil spill occurred in 1997, these are the only observations available at present. Representative observation points of SYNOP and SOUND for the DA in this computation are shown in Fig. 7.10.

Computational results were evaluated only for the ratio defined by the oil particles drifted ashore and the amount of oil particles on the coast of each prefecture. The results are shown in Table 7.9.

Table 7.9 Results of the oil drift prediction at each prefecture

w/o DA						
Prediction Day	1) Jan 2	2) Jan 7	3) Jan 13	4) Jan 19	5) Jan25	Observed
Shimane	1.6	1.6	1.6	1.6	1.6	0.6
Tottori	2.3	21.0	7.7	7.7	7.7	
Hyogo	2.7	2.8	0.8	1.9	1.9	2.9
Kyoto	21.1	22.3	7.8	9.9	10.0	7.3
Fukui	25.2	42.8	46.5	53.5	54.1	37.3
Ishikawa & Toyama	25.5	10.7	18.6	21.2	20.9	44.3
Niigata	15.6	0.2	17.2	4.9	4.6	7.6
other	7.5	0.2	1.5	0.8	0.8	
						(%)
RMSE	11.28	17.90	12.37	12.16	12.38	

3D-VAR						
Prediction Day	1) Jan 2	2) Jan 7	3) Jan 13	4) Jan 19	5) Jan25	Observed
Shimane	1.6	1.6	1.6	1.6	1.6	0.6
Tottori	40.1	18.8	3.5	3.4	3.4	
Hyogo	5.0	3.0	0.7	1.4	1.4	2.9
Kyoto	18.6	22.2	6.9	5.6	5.1	7.3
Fukui	20.0	44.2	40.7	44.9	47.3	37.3
Ishikawa & Toyama	6.9	11.1	21.7	32.7	30.6	44.3
Niigata	6.6	0.3	24.8	10.2	10.2	7.6
other	2.7	0.3	1.8	1.8	2.0	
						(%)
RMSE	24.24	17.40	11.86	6.11	7.31	

4D-VAR						
Prediction Day	1) Jan 2	2) Jan 7	3) Jan 13	4) Jan 19	5) Jan 25	Observed
Shimane	NOT AVAILABLE	1.6	1.5	1.5	1.5	0.6
Tottori		21.5	3.9	3.8	3.8	
Hyogo		3.1	0.6	1.3	1.3	2.9
Kyoto		21.6	7.7	6.1	5.4	7.3
Fukui		41.3	40.0	45.0	47.8	37.3
Ishikawa & Toyama		12.2	21.5	32.3	30.0	44.3
Niigata		0.2	24.0	9.9	10.0	7.6
other		0.1	2.3	1.7	1.8	
						(%)
RMSE		17.37	11.72	6.27	7.63	

It was difficult to evaluate the accuracy from the viewpoint of that ratio, however; therefore, averaged error defined by the RMSE was also calculated and is shown in the last line of Table 7.9.

RMSE was defined as follows:

$$\mathrm{RMSE} = \sqrt{\sum_{1}^{N} \frac{(Ratio_o - Ratio_c)^2}{N}} \tag{7.10}$$

where $Ratio_0$ and $Ratio_c$ are the observed and computed ratio of the oil drifted ashore at each prefecture, respectively, and N is the number of prefectures.

In the case of the forecast without DA, the predicted distribution of the oil that had drifted ashore was scattered by simulation start time. The averaged error defined by RMSE was not much different from the start time of the simulation, but the accuracy was not so fine. On the other hand, the forecast with DA showed gradual improvement as time progressed from the start time in terms of the ratio of oil drifted ashore and averaged error by RMSE.

In particular, the predictions of oil behavior on January 19 and 25 showed good agreement with observations. In the case of applying the DA scheme, the results of 3D-Var and 4D-Var showed a similar tendency. From the results, we concluded that the DA scheme could improve the prediction of oil behavior, and time span to predict such behavior correctly was less than 2 weeks when the DA scheme was applied.

It should be noted that the simulation applying the 4D-Var DA scheme with a starting time of January 2 could not be carried out due to the divergence of the numerical solution.

7.4.2 DA for Estimation of Oil Drifts Using ROMS and WRF (Suzuki et al. 2015)

To evaluate the efficacy of DA for oil drifting simulation using ROMS, a spilled oil drifting simulation with DA for the Nakhodka spill was carried out again. Observations were SYNOP and SOUND, as in the previous simulation. The representative observation points were the ones shown in Fig. 7.10.

The simulation domains for ROMS and WRF are shown in Fig. 7.11. For ROMS, the monthly mean flow rate, salinity, and seawater temperature are given at the open boundary, while the wind stress calculated by WRF is given at the sea surface as an external force. The conditions of the simulation are shown in Table 7.10. The grid size of each computation was set to 5 km. The 5-km mesh was not sufficient to resolve detailed phenomena like the Langmuir circulations (Simecek-Beatty and Lehr 2007). However, we selected this mesh size to shorten the computational time. A number of layers for ROMS and WRF were decided by referring to other calculation examples. Forty layers were selected for ROMS and 17 layers for WRF.

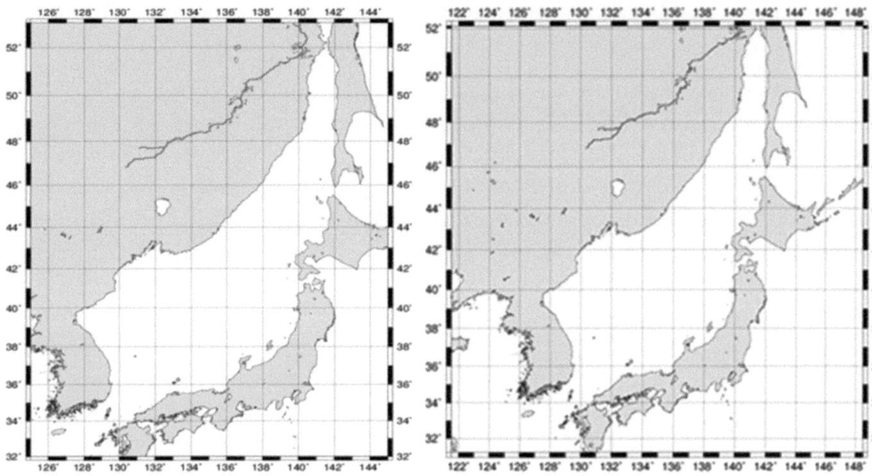

a) ROMS b) WRF

Fig. 7.11 Computational domains for ROMS and WRF

Table 7.10 Computational conditions for the simulation with ROMS and WRF

Model		ROMS	WRF
Domain	Latitude (degree)	32–53	31.2–52.1
	Longitude (degree)	125–145	121.5–148.5
Grid size		5 × 5 km	5 × 5 km
Number of vertical layers		40	17
Time step		Outer:12 s	30 s
		Inner:60 s	
Period		January 1–January 31, 1997	January 1–January 31, 1997
		(November 1–December 31, 1996, as approach run)	
Initial conditions		Results of SODA	NCEP /NCAR global reanalysis products
Boundary conditions		JCOPE2	

The calculation period was set at 30 days, the period of time it should have taken the oil to drift to the coast. The DA scheme we used was the 3D-Var implemented in WRFDA with the cv5 background covariant. For comparison, a simulation without the DA scheme was also computed. Computational conditions including initial/boundary conditions for ROMS and WRF are given in Table 7.10.

7.4.3 Computational Results

The computational results of the spilled oil distributions are shown in Figs. 7.12 and 7.13. In the figures, black dots represent the spilled oil. Figure 7.12 shows the computation without DA. In this computation, the oil stays off the coast of the Fukui and Ishikawa prefectures and does not cross to the Noto Peninsula on January 13. On January 19, the spilled oil crosses the Noto Peninsula and drifts off the coast of Niigata Prefecture. In the end, the oil drifts ashore on the coasts of the Kyoto, Fukui, Ishikawa, and Niigata prefectures.

The computational results of the DA with real observations are shown in Fig. 7.13. In the computation, spilled oil drifts off the coast of the Fukui prefecture on January 7. However, in the observations, the oil drifted off the coast of both Fukui and Kyoto prefectures and crossed the Noto Peninsula on January 13. It did not cross the peninsula in the computation at this time. On the other hand, the drifting oil off the coast of the Tottori prefecture was expressed by the computation. By this time, the spilled oil had drifted ashore to the coasts of the Fukui and Ishikawa prefectures in the computation.

The behavior of spilled oil as time progressed can be observed in the figures. It remains difficult to evaluate whether the computation was correct or not. Therefore, we also evaluated the accuracy from the viewpoint of the quantity of oil that drifted ashore in each prefecture.

The simulation results are shown in Table 7.11. Here, the ratio is defined by the oil particles that have drifted ashore and number of oil particles (of the 8663) in each prefecture. The greatest amount of oil drifted ashore in the area of the

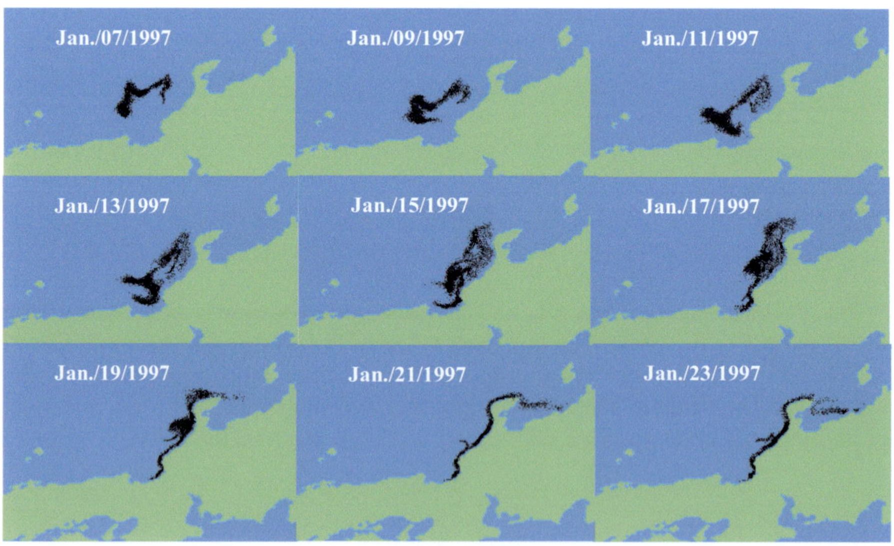

Fig. 7.12 Simulated behavior of drifting oil (without DA)

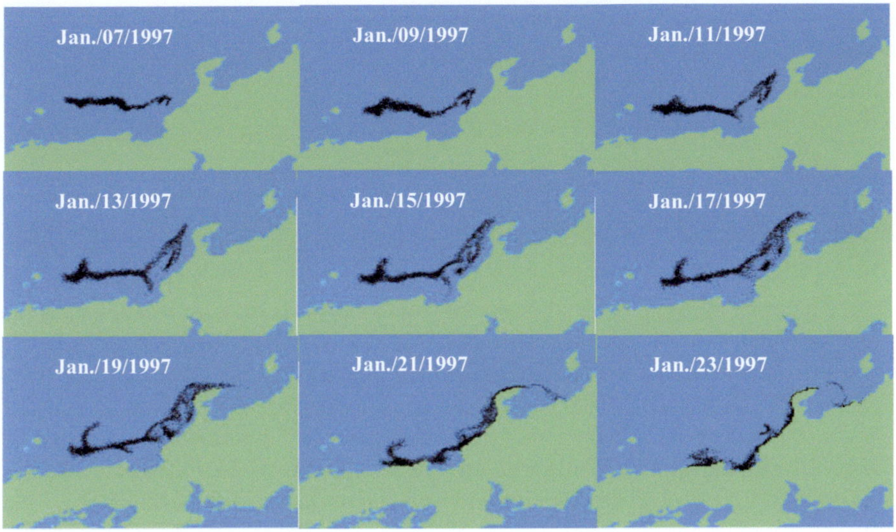

Fig. 7.13 Simulated behavior of drifting oil (with DA: real observations)

Table 7.11 Computed results (amount of oil drifted ashore)

Prefecture	w/o DA	DA (Obs)	Observed
Shimane	0.0	0.0	0.6
Tottori	0.0	0.0	
Hyogo	0.0	0.2	2.9
Kyoto	0.0	24.5	7.3
Fukui	19.3	25.8	37.3
Ishikawa and Toyama	74.6	47.1	44.3
Niigata	6.1	2.5	7.6
			(%)
RMSE	14.7	8.8	

coast of the Ishikawa prefecture in both the observations and computations with and without DA. In case of the simulation without DA, the phenomena of spilled oil drifting ashore at the Hyogo, Kyoto, and Tottori coasts could not be expressed. In contrast, the simulation with DA could express the oil drifted ashore at Hyogo and Kyoto prefectures. The ratio of oil drifted ashore at the Ishikawa Toyama, and Fukui Prefecture was improved by applying the DA scheme. However, the ratio of oil at the Niigata Prefecture was decreased, while that at the Kyoto Prefecture was excessive.

Thus, improvements and drawbacks were observed in the calculation of the distribution of oil drifting ashore at each prefecture by applying the DA scheme. Therefore, the overall averaged accuracy of computational results was evaluated by RMSE, defined in Eq. 7.10. The calculated RMSE are listed in the final line of Table 7.11. From these, we found that the RMSE of simulated results with DA decreased compared to those without DA. This shows that DA can improve the accuracy of the simulation.

7.5 Efficacy of Data Assimilation with SOTAB-II

7.5.1 SOTAB-II

A recent photo of the SOTAB-II is shown in Fig. 7.14; its details were given in the previous chapter.

In this computation, we consider real-time atmospheric observation data from the SOTAB-II for the DA. SOTAB-II is an autonomous buoy robot capable of tracking spilled oil. It is designed to relay not only positional information but also environmental data such as surrounding current velocity and wind velocity to the land station. These positional and/or environmental data can then be used for the DA simulation.

7.5.2 DA Using Data from SOTAB-II

For the actual operation of SOTAB-II, multiple units will be launched into the area of an accident in order to efficiently track the drifting spilled oil. In this simulation, SOTAB-IIs were placed at 500-m intervals in order to surround the spilled oil, as

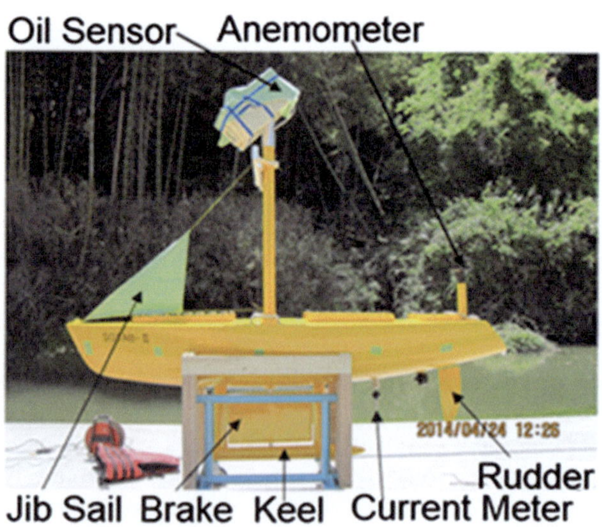

Fig. 7.14 Photo of SOTAB-II

proposed by Senga et. al. (2012). Based on this requirement, 20 SOTAB-IIs were deployed for this accident.

It was first necessary to consider the initial arrangement of the SOTAB-IIs, however, because the heavy oil was spilled from both the sunken hull and the drifted bow of the Nakhodka in the accident. To track the spilled oil effectively, SOTAB-IIs would need to track the bow. On the other hand, the original concept of SOTAB-II was based on oil spills occurring at oil production facilities, like the 2010 Deepwater Horizon accident in the Gulf of Mexico.

As a result, two simulation cases were considered. In the first, the SOTAB-IIs track the bow and spilled oil from the sunken hull; in the second, they track the spilled oil from the sunken hull only. Hereafter, these computations are referred to as "SOTAB-II drifted bow (DB)" and "SOTAB-II sunken hull (SH)," respectively. In the former, 16 SOTAB-IIs (about 80 %) track the Nakhodka bow.

Currently, we don't have a sufficient number of SOTAB-IIs to obtain actual observations. Moreover, SOTAB-II did not exist when the Nakhodka accident occurred. Therefore, the data from SOTAB-II is estimated by interpolating the atmospheric data computed with high precision in place of real observations. Note that it was supposed that the motion of the SOTAB-IIs was the same as the spilled oil, because this is its development concept.

It should also be noted that to evaluate the efficacy of the data from SOTAB-IIs, data of real observations and positional information oil from SOTAB-IIs are not used in these computations although SOTAB-II is planned to measure environmental data and positional information of oil. SOTAB-II can send these data to land station frequently but environmental data every 6 h in the computation of DA.

7.5.3 Computational Results

Computational results for the spilled oil distributions with time progression of the SOTAB-II SH and SOTAB-II DB are shown in Figs. 7.15 and 7.16, respectively. In the figures, black dots represent the spilled oil and yellow dots the SOTAB-IIs.

As discussed above, in the case of the computational results without DA shown in Fig. 7.12, oil stayed off the coast of Fukui and Ishikawa prefectures and did not cross to the Noto Peninsula on January 13. On January 19, spilled oil crossed the Noto Peninsula and drifted off the coast of Niigata Prefecture. In the end, the oil drifted ashore at the coast of Kyoto, Fukui, Ishikawa, and Niigata prefectures. In the case of the computational results of DA with real observations, spilled oil drifted off the coast of Fukui Prefecture on January 7 and did not cross the Noto Peninsula on January 13; drifted oil off the coast of Tottori Prefecture was expressed by the computation, however. By this time, spilled oil had drifted ashore on the coast of Fukui and Ishikawa prefectures, as explained previously.

In case of the computation of SOTAB-II SH shown in Fig. 7.15, the behavior of the oil is similar to the computational results of DA with observations, as shown in the figures from January 7 to 13. The SOTAB-IIs move in a concentrated manner.

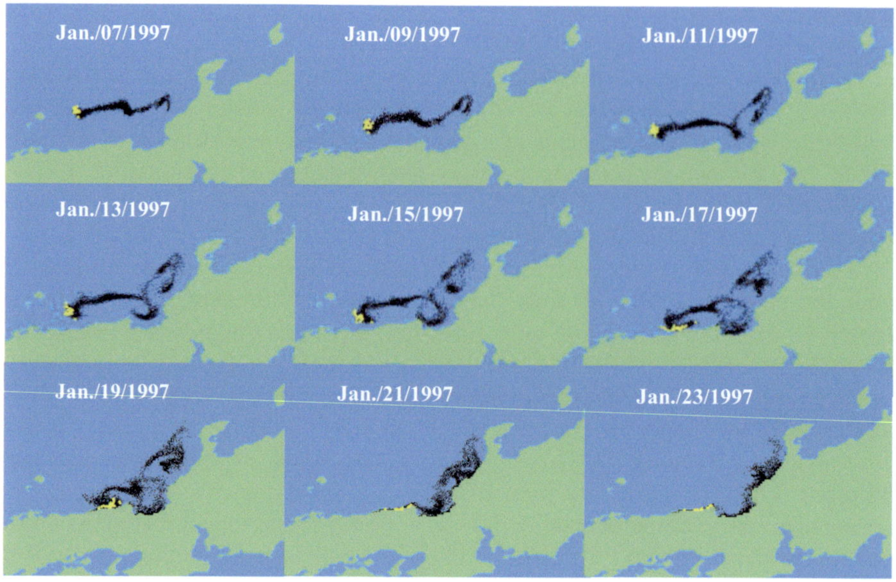

Fig. 7.15 Simulated behavior of drifting oil (with DA: SOTAB-II SH)

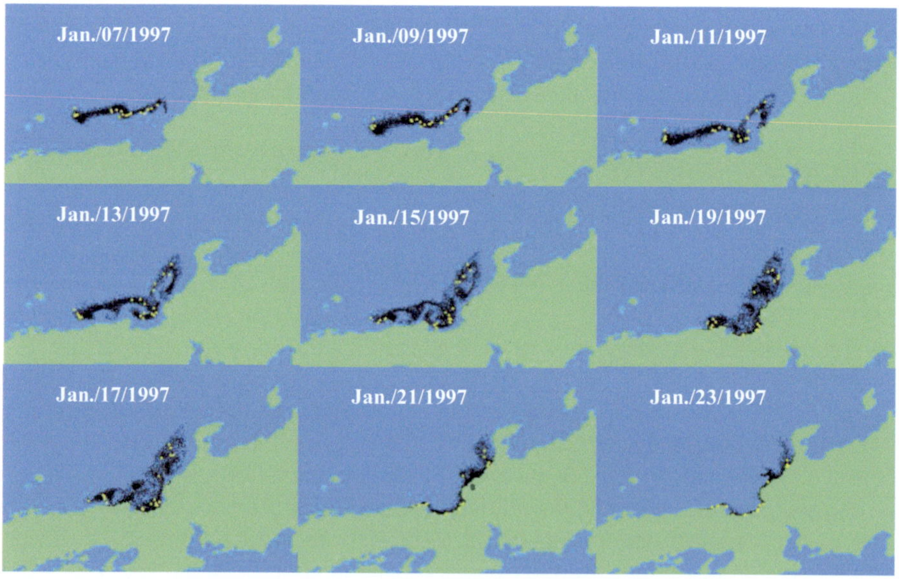

Fig. 7.16 Simulated behavior of drifting oil (with DA: SOTAB-II DB)

This is because SOTAB-II does not track the bow of the Nakhodka. On January 17, oil has drifted ashore to Tottori, Hyogo, Kyoto, and Fukui prefectures. On January 23, much of the oil has drifted ashore to the coast, but no oil crossed the

Noto Peninsula. All SOTAB-IIs drifted ashore to the coast of Hyogo and Kyoto prefectures.

In case of computation of SOTAB-II DB shown in Fig. 7.16, SOTAB-IIs are equally distributed, due to tracking the bow.

Around January 9, we found oil drifting ashore at the Fukui Prefecture. On January 13, oil did not cross the Noto Peninsula in the computation. On the contrary, the westernmost tip of oil was located off the coast of Tottori Prefecture. Around January 15, oil drifted ashore to Kyoto and Fukui prefectures but still did not cross the Noto Peninsula. On January 19, oil drifted ashore to Hyogo, Kyoto, Fukui, and Ishikawa prefectures but again did not drift ashore on the Noto Peninsula in Ishikawa Prefecture. Around January 21, the oil drifted ashore to the Tottori, Hyogo, Kyoto, Fukui, and Ishikawa prefectures. On the other hand, oil did not drift ashore at Niigata prefecture. Finally, all SOTAB-IIs were the equally distributed in the range of the oil that had drifted ashore.

To evaluate these computational results, the ratio of the oil that drifted ashore at each prefecture and RMSE from these ratios was calculated for the overall averaged accuracy, in the same manner as previously. The results are presented in Table 7.12. In the table, DA SH (20) and DA DB (20) indicate "with DA, Track the oil from sunken hull by 20 SOTAB-IIs" and "with DA, Track the oil from sunken hull and drifted bow by 20 SOTAB-IIs," respectively. The RMSE of all simulated results with DA were better than those without DA in comparison to observed data. The case of SOTAB-II DB showed that the smallest error within these simulations, although the oil that drifted ashore to Hyogo and Niigata prefectures, could not be expressed.

It is thought that the locations of the SOTAB-IIs in case of SOTAB-II SH were so narrow that the DA algorithm did not work well. On the other hand, the spacing of the SOTAB-IIs was relatively wide in the case of the SOTAB-II DB, and here the DA scheme worked effectively. In other words, it is better if the observations include a wide range in the computational domain. This shows that if SOTAB-IIs were arranged properly, they would be effective in improving numerical accuracy when DA was applied using their data.

Table 7.12 Computed results (amount of oil drifted ashore)

Prefecture	w/o DA	DA (Obs)	DA SH(20)	DA DB(20)	Observed
Shimane	0.0	0.0	0.0	0.0	0.6
Tottori	0.0	0.0	0.0	0.0	
Hyogo	0.0	0.2	0.4	0.0	2.9
Kyoto	0.0	24.5	29.3	12.3	7.3
Fukui	19.3	25.8	38.1	45.8	37.3
Ishikawa and Toyama	74.6	47.1	32.3	41.9	44.3
Niigata	6.1	2.5	0.0	0.0	7.6
					(%)
RMSE	14.7	8.8	10.7	5.3	

7.6 Effect of Number of SOTAB-IIs

As already mentioned, the amount of oil spilled by the Nakhodka oil would have required 20 SOTAB-IIs. This number has not been confirmed, however. In this section, 10 and 30 SOTAB-IIs are considered to confirm the required number of SOTAB-IIs for the accident.

In both cases, about 80 % of the SOTAB-IIs track the bow of the Nakhodka, and about 20 % track the oil spilled from the hull, as in the SOTAB-II DB computation in the previous section.

7.6.1 Computational Results

The time histories of the computational results with 10 SOTAB-IIs are shown in Fig. 7.17, and with 30 SOTAB-IIs in Fig. 7.18. In the figures, black dots and yellow dots represent the spilled oil and SOTAB-IIs, respectively, as before.

There were no large differences between the conditions in terms of overall computational results. In the details, however, in the computation with 10 SOTAB-IIs, SOTAB-II was no longer present in the eastern part of the spilled oil on January 7, compared to the case of SOTAB-II DB (20). This meant the prediction accuracy at the leading point (eastern part) of the drifting oil could not be expected. On the other hand, in case of the computation with 30 SOTAB-IIs, although the density of

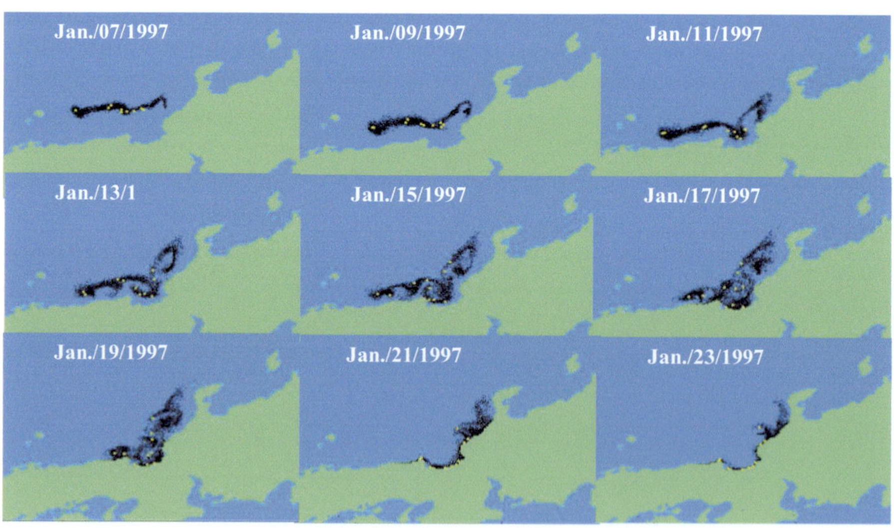

Fig. 7.17 Simulated behavior of drifting oil (with DA: SOTAB-II DB [10 SOTAB-IIs])

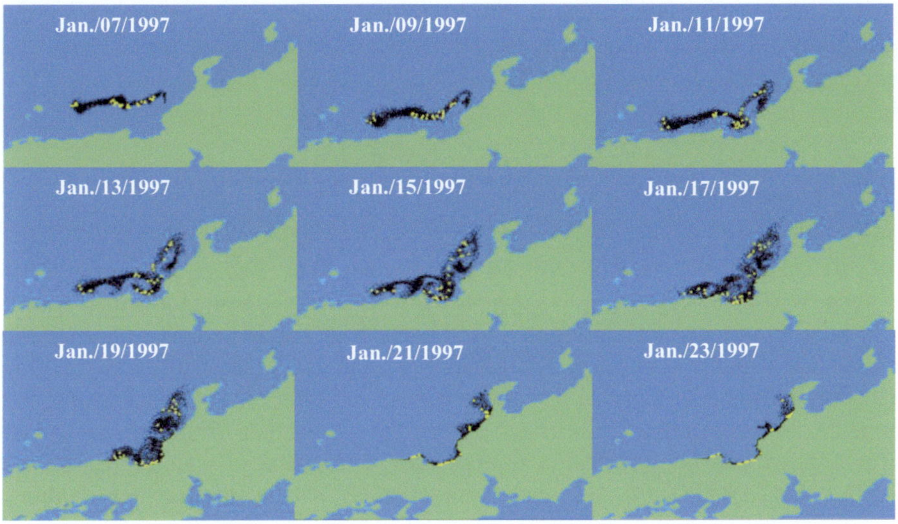

Fig. 7.18 Simulated behavior of drifting oil (with DA: SOTAB-II DB [30 SOTAB-IIs])

Table 7.13 Computed results (amount of oil drifted ashore)

Prefecture	Without DA	DA (Obs)	DA SH(20)	DA DB(20)	DA DB(10)	DA DB(30)	Observed
Shimane	0.0	0.0	0.0	0.0	0.0	0.0	0.6
Tottori	0.0	0.0	0.0	0.0	0.0	0.0	
Hyogo	0.0	0.2	0.4	0.0	0.0	0.0	2.9
Kyoto	0.0	24.5	29.3	12.3	11.8	10.5	7.3
Fukui	19.3	25.8	38.1	45.8	48.1	46.8	37.3
Ishikawa and Toyama	74.6	47.1	32.3	41.9	40.0	42.7	44.3
Niigata	6.1	2.5	0.0	0.0	0.0	0.0	7.6
							(%)
RMSE	14.7	8.8	10.7	5.3	6.1	5.3	

SOTAB-II units was lower, some units were present near the leading point of the drifting oil.

As before, the ratio of the amount of oil drifting ashore was calculated, and RMSE (Eq. 7.10) was also calculated as the overall averaged error evaluation. The calculated results are shown in Table 7.13. The DA scheme (3D-Var [cv5]) was applied to both DA DB (10) and DA DB (30).

Based on the computational results of "DA DB (20)," in which 20 SOTAB-IIs tracked the oil from the sunken hull and drifted bow, "DA DB (10)" showed similar tendencies, but the amount of the oil drifted ashore at the Fukui Prefecture was slightly larger, and that at Ishikawa and Toyama prefectures was slightly smaller. As a result, overall averaged error based on RMSE was worse.

In contrast, the application of 30 SOTAB-IIs improved computation of the amount of oil drifted ashore at the Fukui, Ishikawa, and Toyama prefectures. The averaged error in the case of "DA DB (20)" and "DA DB (30)" was unchanged.

7.7 Conclusion

In this chapter, a spilled oil drifting simulation was carried out using the 1997 Nakhodka oil spill in the Sea of Japan as a test case for the simulation. To improve the accuracy of the simulation, a DA scheme was applied for the atmospheric simulation using observations. Before applying the DA scheme to the oil drifting simulation, numerical experiments were carried out to select the data assimilation scheme. Subsequently, the efficacy of the DA when data from the SOTAB-II was applied was investigated to forecast the phenomena of oil drift on the sea surface. SOTAB-II is still under development; interpolated data from high-accuracy simulated wind data was treated as data from SOTAB-IIs. The effect of number of SOTAB-IIs was also investigated.

Our results can be summarized as follows:

1. The motion of the heavy oil spilled from the Nakhodka could be computed with our method.
2. The amounts of oil drifted ashore to each prefecture were calculated.
3. Within DA schemes implemented in WRFDA, we tested 3D-Var and 4D-Var with the background error covariance option cv3 and cv5. We selected 3D-Var (cv5) from the viewpoint of computational accuracy and load.
4. An averaged accuracy of computation was defined by the amount of the oil drifted ashore at each prefecture using RMSE. Based on this, the RMSE of each case was calculated. Results demonstrated that the accuracy of the prediction was improved by applying the DA scheme.
5. The time span to predict the spilled oil drifting behavior correctly is less than 2 weeks when the DA scheme is applied.
6. The data from SOTAB-II was effective for DA computation.
7. The appropriate number of SOTAB-IIs is based on the scale of the accident (total amount of spilled oil).
8. In the case of the Nakhodka oil spill, the appropriate number of SOTAB-II was 20.
9. Observations for the DA were better when they existed in a wide range of the computational domain.

However, from the viewpoint of the motion of the spilled oil over time and the amount of oil drifting ashore at each prefecture, it will be necessary to carry out further improvements in numerical accuracy. To improve accuracy, we must test the grid dependency carefully and introduce DA to the ocean model. In the future, this simulation will also be performed by incorporating real-time measured data obtained from the SOTAB-II.

Acknowledgments The author of this chapter gratefully acknowledges the work of the past and present members of Kato laboratory, Department of Naval Architecture and Ocean Engineering, Osaka University, especially, Mr. Youhei Mukumoto, Mr. Yuichi Miyajima, Mr. Kousuke Iwai, and Mr. Satoaki Tsutsukawa.

References

Fairall CW et al (1996) Bulk parameterization of air-sea fluxes for tropical ocean-global atmosphere coupled-ocean atmosphere response experiment. J Geophys Res 101(C2):3747–3764

Fay JA (1971) Physical processes in the spread of oil on a water surface. In: International oil spill conference proceedings, Vol 1971. No. 1. pp 463–467

Ishii J et al (1998) In: Heavy oil pollution: for tomorrow can "Nakhodka" change Japan? Ocean Engineering Research, Tokyo

Mackay D, Zagorski W (1982) Studies of water-in-oil emulsions, Environment Canada Manuscript Report EE-34

Matsuzaki Y, Fujita I (2012) A numerical simulation method of oil transport considering oil spreading, Technical note of the Port and airport research institute No.1225

Senga H et al (2012) Verification experiments of sail control effects on tracking oil spill oceans. IEEE May 21–24, Proc. (cd-rom)

Simecek-Beatty D, Lehr J (2007) Trajectory modeling of marine oil spills. In: Wang Z, Stout SA (eds) Oil spill environmental forensics fingerprinting and source identification. Academic Press, Boston, pp 405–418

Smagorinsky J (1963) General circulation experiments with the primitive equations 1. The basic experiment. Monthly Weather Rev 91(3)

Suzuki S et al (2015) Study on spilled oil drifting simulation applying data assimilation with SOTAB-II. In: Proceedings of international society of offshore polar engineer conference (ISOPE2015) Hawaii, USA. June 21–26, 2015(cd-rom)

Tsutsukawa S et al (2012) Efficacy evaluation of data assimilation for simulation method of spilled oil drifting. In: Proceedings of 5th PAAMES and AMEC2012. Taiwan, Paper No. SEPAS-05

Chapter 8
Conclusions

Naomi Kato

Keywords SOTAB project • Conclusions • Future work

This book addressed the whole view of the SOTAB project initiated by the Deepwater Horizon (DWH) oil spill accident off Louisiana in the Gulf of Mexico (GOM) in 2010, which is considered to be the largest in the history of the petroleum industry. It was conducted from 2011FY to 2015FY with the support of a Grant-in-Aid for Scientific Research (S) of the Japan Society for the Promotion of Science. The conclusions from Chaps. 2, 3, 4, 5, 6, and 7 can be summarized as follows.

In Chap. 2, work on the change in the structure of subsurface plumes after the DWH oil spill accident was reviewed after showing major marine-based oil spill and gas leak accidents in the world. It addressed the concentrations of dissolved hydrocarbon gases between 1,000 and 1,300 m depth, and the formation of a residual microbial community containing methanotrophic bacteria after a large quantity of oxygen was respired primarily by methanotrophs within about 120 days from the onset of release of methane gas from the DWH wellhead. It also reviewed the work on the behavior of surface oil slicks after the DWH oil spill accident from the viewpoint of the effects of the ocean circulation patterns of the GOM Loop Current (LC) system on the advection of the oil discharged during the DWH accident. It showed that the merging of three cyclonic eddies along the LC's northern margin finally formed a larger and stronger cyclonic eddy and that this larger cyclonic eddy controlled the motion of the oil/dispersant mixture into the deep water.

In Chap. 3, the effectiveness of a progressive depth control scheme consisting of time estimation depth control and PID depth control was evaluated numerically and experimentally to shorten the time for descending to the seabed and to avoid collision with the seabed. SOTAB-I was able to collect dissolved substances in seawater, such as hydrocarbons, from the early deployments of SOTAB-I in the shallow water of the GOM in the USA. The measurements not only of vertical water column distribution of oceanographic data, such as temperature, salinity, and

N. Kato (✉)
Department of Naval Architecture and Ocean Engineering, Osaka University, 2-1, Yamada-oka, Suita 565-0871, Osaka, Japan
e-mail: kato@naoe.eng.osaka-u.ac.jp

© Springer Japan 2017
N. Kato (ed.), *Applications to Marine Disaster Prevention*,
DOI 10.1007/978-4-431-55991-7_8

density, but also of a high-resolution profile of water currents by SOTAB-I have been achieved through deepwater experiments conducted in Toyama Bay in Japan.

In Chap. 4, the effectiveness of a clustering-based decision-making algorithm for sailing the autonomous surface vehicle (ASV) based on a complete time history of the scanned area around the vehicle by the oil-detecting sensor was clarified in reacting to changing wind conditions as well as dynamic relative positioning of SOTAB-II to oil slicks. However, sea experiments using actual oil slicks on the sea surface in Japan could not be carried out because of regulations banning the use of actual oil along the Japanese coastline.

In Chap. 5, it was clarified that the simple oil tracking model for tracking oil from the seabed in deep water based on Lagrangian approaches was valid if droplet breakup and coalescence did not need to be considered and that the standard hydrate shell model governed by thermophysical stability gave a reasonable result for predicting large-scale motion. However, the physics of rising methane gas/hydrate was strongly affected by the empirical parameters and the small-scale modeling.

In Chap. 6, the interaction between liquid and gas in the multiphase flow, especially between dispersed gas bubbles and oil droplets in three-phase blowout flow, was discussed through numerical simulations where the mass and momentum balance equations were solved for each phase and the interactions were considered with the Euler–Euler approach, showing that such an interaction affects the rising velocity of each phase.

In Chap. 7, the data assimilation scheme that incorporates measured data into the numerical simulation was discussed to increase the prediction accuracy of surface oil drifting assuming meteorological data from the multiple ASV SOTAB-IIs. It was successful in showing the applicability of its scheme through numerical simulation.

For future work, the following items should be further investigated:

1. Multiple SOTAB-IIs

   It is necessary to deploy multiple SOTAB-IIs to track widely spread oil slicks by communicating with each other and a land station.
2. Innovative system for marine disaster prevention using a data assimilation scheme:

   (a) Chapter 7 showed the effectiveness of the data assimilation scheme incorporating the artificially measured data from multiple SOTAB-IIs into the numerical simulation. It is feasible to achieve prediction of oil drifting in a wide area in real time using the data assimilation scheme if the first item for future work can be achieved.

   (b) From this sense, the underwater autonomous robot SOTAB-I with the functions of measuring vertical profiles not only of water currents, water temperature, and water salinity around the oil and gas spill region using ADCP and CTD but also of dissolved gas and volatile organic compounds as shown in Chap. 3 can provide the simulation with underwater behavior of the oil and gas blowout with the measured data to raise its accuracy.

   We carried out exploration of a deepwater methane plume along Toyama Trough off Joetsu where a methane plume seeped from the seabed (Matsumoto et al. 2009) on September 2–4, 2015, using SOTAB-I equipped with UMS measuring dissolved gas. We prepared the simulation program for the methane plume movement in underwater currents explained in Chap. 5 and that for movement of SOTAB-I in water currents to get the optimum path planning of SOTAB-I in obtaining the maximum data inside the plume during its descending and ascending operations, as explained in Chap. 3. The vertical profiles of water currents, water temperature, and water salinity can be given by the first dive of SOTAB-I to the seabed near the methane seepage point that was identified using an echo sounder of the mother ship (Chiba et al. 2015). Then, the optimum path and starting point of the exploration of the methane plume can be determined assuming the amount of seepage of methane gas from the seabed. However, a fatal fault was found after the first dive near the methane seepage point; a water leak inside the UMS was found. We had to abandon the exploration at this time. We should seek a further chance for the exploration.

Toyama Trough is located on the extended line from Itoigawa-Shizuoka Tectonic Line well known as active fault zone. A cruise report by JAMSTEC (Sato et al. 2008) explained the investigation of the geological structure of Toyama Trough and showed that layers of methane hydrate under the seabed are formed in the neighborhood of faults. It is valuable to furthermore investigate the relationship between seismic activities along Toyama Trough and seepage of methane gas from the viewpoint of marine disaster prevention.

References

Chiba H et al (2015) Identification of location of methane seepage from seabed using echo sounder and GPS. Conference Proceeding of Japan Institute of Navigation K. 133–15, pp 137–140 (in Japanese)

Matsumoto R et al (2009) Formation and collapse of gas hydrate deposits in a high methane flux area of the Joetsu Basin. East Margin Jpn Sea J Geogr 118:43–71 (in Japanese with abstract in English)

Sato M et al (2008) Investigation of geologic structures over methane hydrate system under the seabed off Joetsu. Blue Earth'08, JAMSTEC, Yokohama, 13–14 Mar 2008 (in Japanese)

MIX
Papier aus verantwortungsvollen Quellen
Paper from responsible sources
FSC® C105338

If you have any concerns about our products,
you can contact us on
ProductSafety@springernature.com

In case Publisher is established outside the EU,
the EU authorized representative is:
Springer Nature Customer Service Center GmbH
Europaplatz 3, 69115 Heidelberg, Germany

Printed by Libri Plureos GmbH
in Hamburg, Germany